Integrated Circuit Fabrication

Integrated Circuit Fabrication

Dr. Kumar Shubham
Associate Professor
Delhi Technical Campus
Greater Noida, Uttar Pradesh
India

Ankaj Gupta
Assistant Professor
Delhi Technical Campus
Greater Noida, Uttar Pradesh
India

CRC Press
Taylor & Francis Group
Boca Raton London New York

CRC Press is an imprint of the
Taylor & Francis Group, an **informa** business

First published 2021
by CRC Press
2 Park Square, Milton Park, Abingdon, Oxon, OX14 4RN

and by CRC Press
6000 Broken Sound Parkway NW, Suite 300, Boca Raton, FL 33487-2742

© 2021 Manakin Press Pvt. Ltd.

CRC Press is an imprint of Informa UK Limited

The rights of Kumar Shubham and Ankaj Gupta to be identified as the authors of this work has been asserted by them in accordance with sections 77 and 78 of the Copyright, Designs and Patents Act 1988.

Reasonable efforts have been made to publish reliable data and information, but the author and publisher cannot assume responsibility for the validity of all materials or the consequences of their use. The authors and publishers have attempted to trace the copyright holders of all material reproduced in this publication and apologize to copyright holders if permission to publish in this form has not been obtained. If any copyright material has not been acknowledged please write and let us know so we may rectify in any future reprint.

All rights reserved. No part of this book may be reprinted or reproduced or utilised in any form or by any electronic, mechanical, or other means, now known or hereafter invented, including photocopying and recording, or in any information storage or retrieval system, without permission in writing from the publishers.

For permission to photocopy or use material electronically from this work, access www.copyright.com or contact the Copyright Clearance Center, Inc. (CCC), 222 Rosewood Drive, Danvers, MA 01923, 978-750-8400. For works that are not available on CCC please contact mpkbookspermissions@tandf.co.uk

Trademark notice: Product or corporate names may be trademarks or registered trademarks, and are used only for identification and explanation without intent to infringe.

Print edition not for sale in South Asia (India, Sri Lanka, Nepal, Bangladesh, Pakistan or Bhutan).

British Library Cataloguing-in-Publication Data
A catalogue record for this book is available from the British Library

Library of Congress Cataloging-in-Publication Data
A catalog record has been requested

ISBN: 978-1-032-01429-6 (hbk)
ISBN: 978-1-003-17858-3 (ebk)

To Our Parents
and
Family

Brief Contents

1. Introduction to Silicon Wafer Processing — 1–36
2. Epitaxy — 37–74
3. Oxidation — 75–102
4. Lithography — 103–138
5. Etching — 139–166
6. Diffusion — 167–206
7. Ion Implantation — 207–240
8. Film Deposition: Dielectric, Polysilicon and Metallization — 241–278
9. Packaging — 279–294
10. VLSI Process Integration — 295–324
    Appendix — 325–332
    Index — 333–336

Detail Contents

1. Introduction to Silicon Wafer Processing — 1–36
  1.1 Introduction — 1
  1.2 VLSI Generations — 2
  1.3 Clean Room — 5
  1.4 Semiconductor Materials — 8
  1.5 Crystal Structure — 10
  1.6 Crystal Defects — 13
  1.7 Si properties & its Purification — 18
  1.8 Single Crystal Si Manufacture — 20
      1.8.1 Czochralski Crystal Growth Technique — 20
      1.8.2 Float zone Technique (FZ Technique) — 26
  1.9 Silicon Shaping — 27
  1.10 Wafer Processing Considerations — 32
  1.11 Summary — 33
      Problems — 33
      References — 35

2. Epitaxy — 37–74
  2.1 Introduction — 37
  2.2 Liquid Phase Epitaxy — 39
  2.3 Vapor Phase Epitaxy/Chemical Vapor Deposition — 42
      2.3.1 Growth Model and Theoretical Treatment — 44
      2.3.2 Growth Chemistry — 46
      2.3.3 Doping — 48
      2.3.4 Reactors — 49
  2.4 Defects — 50
  2.5 Technical Issues for Si Epitaxy by CVD — 52
      2.5.1 Uniformity/Quality — 52
      2.5.2 Buried Layer Pattern Transfer — 53
  2.6 Autodoping — 58
  2.7 Selective Epitaxy — 59
  2.8 Low Temperature Epitaxy — 60
  2.9 Physical Vapor Deposition (PVD) — 61
      2.9.1 Molecular Beam Epitaxy (MBE) — 61
  2.10 Silcon-on-Insulator (SOI) — 66

2.11	Silicon on Sapphire (SOS)	67
2.12	Silicon on SiO$_2$	68
2.13	Summary	68
	Problems	69
	References	70

3. Oxidation 75–102

3.1	Introduction	75
3.2	Growth and Kinetics	78
	3.2.1 Dry Oxidation	79
	3.2.2 Wet Oxidation	80
3.3	Growth Rate of Silicon Oxide Layer	82
3.4	Impurities effect on the Oxidation Rate	87
3.5	Oxide Properties	89
3.6	Oxide Charges	90
3.7	Oxidation Techniques	92
3.8	Oxide Thickness Measurement	92
3.9	Oxide Furnaces	95
3.10	Summary	98
	Problems	98
	Reference	99

4. Lithography 103–138

4.1	Introduction	103
4.2	Optical Lithography	105
4.3	Contact Optical Lithography	106
4.4	Proximity Optical Lithography	106
4.5	Projection Optical Lithography	107
4.6	Masks	112
4.7	Photomask Fabrication	114
4.8	Phase Shifting Mask	115
4.9	Photoresist	116
4.10	Pattern Transfer	119
4.11	Particle-Based Lithography	122
	4.11.1 Electron Beam Lithography	122
	4.11.2 Electron-Matter Interaction	124
4.12	Ion Beam Lithography	127
4.13	Ultra Violet Lithography	129
4.14	X-Ray Lithography	130
4.15	Comparison of Lithographic Techniques	132
4.16	Summary	133
	Problems	134
	References	139

5. Etching   139–166
  5.1 Introduction   139
  5.2 Etch Parameters   139
  5.3 Wet Etching Process   141
  5.4 Silicon Etching   143
  5.5 Silicon Dioxide Etching   145
  5.6 Aluminum Etching   146
  5.7 Dry Etching Process   147
  5.8 Plasma Etching Process   147
       5.8.1 Plasma Chemical Etching Process   150
       5.8.2 Sputter Etching Process   151
       5.8.3 Reactive Ion Etching (RIE) Process   152
  5.9 Inductive coupled Plasma Etching (ICP)   153
  5.10 Advantages and Disadvantages of Dry Etching
       (Plasma Etching) and Wet Etching   154
  5.11 Examples of Etching Reactions   154
  5.12 Liftoff   157
  5.13 Summary   159
       Problems   159
       References   160

6. Diffusion   167–206
  6.1 Introduction   167
  6.2 Atomic Mechanisms of Diffusion   168
       6.2.1 Substitutional Diffusion   168
       6.2.2 Interstitial Diffusion   169
  6.3 Fick's Laws of Diffusion   171
  6.4 Diffusion Profiles   172
       6.4.1 Constant Source Concentration Distribution   173
       6.4.2 Limited Source Diffusion or Gaussian Diffusion   175
  6.5 Dual Diffusion Process   177
       6.5.1 Intrinsic & Extrinsic Diffusion   179
       6.5.2 Diffusivity of Antimony in Silicon   182
       6.5.3 Diffusivity of Arsenic in Silicon   182
       6.5.4 Diffusivity of Boron in Silicon   183
       6.5.5 Diffusivity of Phosphorus in Silicon   185
  6.6 Emitter Push Effect   186
  6.7 Field-Aided Diffusion   188
  6.8 Diffusion Systems   189
  6.9 Oxide Masking   193
  6.10 Impurity Redistribution During Oxide Growth   195
  6.11 Lateral Diffusion   196

6.12	Diffusion in Polysilicon	197
6.13	Measurement Techniques	198
	6.13.1 Staining	198
	6.13.2 Capacitance-Voltage Plotting (C-V)	199
	6.13.3 Four Point Probe (FPP)	200
	6.13.4 Secondary Ion Mass Spectroscopy (SIMS)	201
	6.13.5 Spreading Resistance Probe (SRP)	201
6.14	Summary	202
	Problems	202
	References	202

7. Ion Implantation   207–240

7.1	Introduction	207
7.2	Ion Implanter	209
	7.2.1 Gas System	210
	7.2.2 Electrical System	210
	7.2.3 Vacuum System	210
	7.2.4 Control System	210
	7.2.5 Beam Line System	210
7.3	Ion Implant Stop Mechanism	213
7.4	Range and Straggle of Ion Implant	217
7.5	Thickness of Masking	220
7.6	Doping Profile of Ion Implant	222
7.7	Annealing	223
	7.7.1 Furnace Annealing	224
	7.7.2 Rapid Thermal Annealing (RTA)	226
7.8	Shallow Junction Formation	228
	7.8.1 Low Energy Implantation	229
	7.8.2 Tilted Ion Beam	229
	7.8.3 Implanted Silicides and Polysilicon	230
7.9	High Energy Implantation	231
7.10	Buried Insulator	232
7.11	Summary	233
	Problems	234
	References	234

8. Film Deposition: Dielectric, Polysilicon and Metallization   241–2278

8.1	Introduction	241
8.2	Physical Vapor Deposition (PVD)	242
	8.2.1 Evaporation	243

8.2.2 Sputtering	244
8.3 Chemical Vapor Deposition (CVD)	245
8.4 Silicon Dioxide	249
8.5 Silicon Nitride	253
8.5.1 Locos Methods	254
8.6 Polysilicon	256
8.7 Metallization	257
8.8 Metallization Application in VLSI	259
8.9 Mettalization Choices	260
8.10 Copper Metallization	261
8.11 Aluminium Metallization	262
8.12 Metallization Processes	265
8.13 Deposition Methods	265
8.14 Deposition Apparatus	266
8.15 Liftoff Process	268
8.16 Multilevel Metallization	269
8.17 Characteristics of Metal Thin Film	271
8.18 Summary	275
Problems	275
References	276

9. Packaging 279–294
9.1 Introduction	279
9.2 Package Types	280
9.3 Packaging Design Considerations	283
9.4 Integrated Circuit Package	284
9.5 VLSI Assembly Technologies	287
9.6 Yield	291
9.7 Summary	293
Problems	293
References	293

10. VLSI Process Integration 295–324
10.1 Introduction	296
10.2 Fundamental Considerations for IC Processing	298
10.3 NMOS IC Technology	298
10.4 CMOS IC Technology	301
10.4.1 N-Well Process	301
10.4.2 P-Well Process	305
10.4.3 Twin Tub Process	306
10.5 Bipolar IC Technology	307

10.6	Bi-CMOS Technology	309
10.7	Bi-CMOS Fabrication	309
10.8	FinFET	312
10.9	Monolithic and Hybrid Integrated Circuits	315
10.10	IC Fabrication / Manufacturing	316
10.11	Fabrication Facilities	317
10.12	Summary	322
	Problems	322
	References	322

Appendix   **325–332**

Index   **333–336**

Preface

The book for students in engineering and technology curricula. It is a comprehensive treatment of integrated circuit fabrication. This text will be very useful to every student and practicing professional dealing with fabrication process. It covers theoretical and practical aspects of all major steps in the fabrication sequence. This book can be used conveniently in a semester length course on integrated circuit fabrication. This text can also serve as a reference for practicing engineer and scientist in the semiconductor industry.

IC Fabrications are ever demanding of technology in rapidly growing industry where growth opportunities are numerous. A recent survey shows that integrated circuit currently outnumbers humans in UK, USA, India and China. The spectacular advances in the development and application of integrated circuit technology have led to the emergence of microelectronic process engineering as an independent discipline.

Integrated circuit fabrication text books typically divide the fabrication sequence into a number of unit processes that are repeated to form the integrated circuit. Most students have difficulty recalling all of the background material. They have seen it once, two or three years and many final exams ago. The effect is to give the book a analysis flavor: a number of loosely related topics each with its own background material. It is vital that this fundamental material be re-established before students take up new material.

Manuscript Organization

Chapter 1 presents brief historical overview of major semiconductor devices and vital technological development, as well as introduction to basic fabrication step. Introduces crystal growth and process to obtain single crystal of silicon.

Chapter 2 deals with epitaxy. The purpose of epitaxy is to grow a silicon layer of uniform thickness and accurately controlled electrical properties and so to provide a perfect substrate for the subsequent device processing.

Chapter 3 presents silicon oxidation. It refers to the conversion of the silicon wafer to silicon oxide (SiO_2 or more generally $SiOx$). The ability of Si to form an oxide layer is very important since this is one of the reasons for choosing Si over Ge.

Chapter 4 discuss lithography. It is the process of transferring patterns of geometric shapes in a mask to a thin layer of radiation-sensitive material (called resist) covering the surface of a semiconductor wafer.

Chapter 5 explain the process of etching that explain dry and wet etching process of different materials.

Chapter 6 refers diffusion to the entire process of adding a dopant to the surface of wafer at high temperature.

Chapter 7 discuss ion-implantation which replaces diffusion process in IC fabrication for reliable and reproducable doping.

Chapter 8 presents Film Deposition which contains two part Dielectric film deposition and metallization. It contains SiO_2 and Si_3N_4 film deposition which act as passivative material in IC. Metallization refers to the metal layers that electrically interconnect the various device structures fabricated on the silicon substrate. Thin-film aluminum is the most widely used material for metallization, and is said to be the third major ingredient for IC fabrication, with the other two being silicon and SiO_2.

Chapter 9 deals with packaging. It is final stage of IC fabrication. In this the tiny block of semiconducting material is encapsulated in a supporting case that prevents physical damage and corrosion.

Chapter 10 is all about VLSI process integration. It presents fundamentals of integrating silicon processing steps to create silicon devices.

In this text over 400 references have been cited, of which 30 percent were published in last ten years, and over 200 technical illustrations are included, of which 45 percent are new. The problems at the end of each chapter form an integral part of the development of the topic. We have also attempted in each chapter to conclude some discussion of future trends. This book is organized somewhat differently than other texts on this general topic.

Suggestions for further improvements of the book will be gratefully acknowledge.

Authors

1

Introduction to Silicon Wafer Processing

1.1 INTRODUCTION

Designing a complex electronic machine of compact size like a laptop or mobile, it is always desired and necessary to increase the number of components involved in order to make technical advanced. The logic operation parts of the machines are conducted through integrated circuits made of semiconductor material. The monolithic integrated circuit placed the previously separated diodes, transistors, resistors, capacitors and all the connecting wiring onto a single crystal (or 'chip'). The monolithic integrated circuit was fated to be invented as two inventors who were unaware of each others activities, invented almost identical integrated circuits or ICs at nearly the same time.

Jack Kilby, an engineer from Texas Instruments in 1958 with a background in ceramic-based silk screen circuit boards and transistor-based hearing aids had similar idea of making a whole circuit on a single chip as of research engineer Robert Noyce who had co-founded the Fairchild Semiconductor Corporation in 1957.

What we didn't realize then was the integrated circuit would reduce the cost of electronic functions by a factor of a million to one, nothing had ever done that for anything before" – Jack Kilby

In 1961 the first commercially available integrated circuits came from the Fairchild Semiconductor Corporation.

Jack Kilbe   Robert Noyce

Fig. 1.1 Photo image of (a) Jack Kilbe (b) Robert Noyce

Jack Kilby holds patents on more than sixty inventions and is also well known as the inventor of the portable calculator (1967), awarded the National Medal of Science in 1970. Robert Noyce, with sixteen patents to his name, founded Intel, the company responsible for the invention of the microprocessor, in 1968. The invention of the integrated circuit by both men stands historically as one of the most important innovations of mankind as almost all modern products use chip technology. Kilby used Germanium and Noyce used silicon for the semiconductor material.

All computers then started to be made using chips instead of assembling the individual transistors and their accompanying parts. Texas Instruments first used the chips in Air Force computers and the Minuteman Missile in 1962. They later used the chips to produce the first electronic portable calculators. The first IC had only one transistor, three resistors, and one capacitor having a size of an adult's pinkie finger. Today an IC smaller than a penny can hold more than 1 billion transistors.

The advantages of integrated circuits are as follows

1. Small in size due to the reduced device dimension
2. Low weight due to very small size
3. Low power requirement due to lower dimension and lower threshold power requirement
4. Low cost due to large-scale production and cheap material
5. High reliability due to the absence of a solder joint
6. Provide facilitation to integrate large number of devices and components.
7. Improves the device performance even at high-frequency region

The disadvantages of integrated circuits are as follows

1. IC resistors have a limited range
2. Due to bulky size generally inductors (L) cannot be formed using IC
3. Transformers cannot be formed using IC.

1.2 VLSI GENERATIONS

Historically, the first semiconductor IC chips held one transistor with three resistors and one capacitor. Advancement of technology enabled us to add more and more number of transistors.

1.2 VLSI Generations

The first to arrive was Small-Scale Integration (SSI), then improvements in technique led to devices with millions to billions of logic gates–Very Large-Scale Integration (VLSI).

Present day's microprocessors have millions of logic gates and transistors. Intel co-founder, Gordon E. Moore, in 1965 published a paper on the future projection of IC technology.

Moore's Law is responsible for "smaller, compact, cheaper and more efficient IC". Gordon Moore's empirical relationship is cited in a number of forms, but its essential thesis is that the numbers of transistors that can be manufactured on a single die will double every 18 months.

Wickes (1969) in his paper categorizes between SSI, MSI and LSI by the number of logic gates implemented on single chip where single equivalent logic gate is taken as the fundamental building block. On this basis a SSI circuit is one which has 1~10 equivalent logic gates, an MSI circuit is one which has 10~100 logic gates and LSI circuit is one which has more than 100 logic gates. (e.g., random access bipolar memory modules have approximately 500 equivalent gates and other advanced modules are expected to have four times the number.) After the success of LSI, the era of Very Large Scale Integration (VLSI), Extra Large Scale Integration (ELSI), Ultra Large Scale Integration (ULSI), etc has begun with having the ability to perform a very complex logic function, or a large number of simple logic functions in very short time.

The first generation integrated circuits contained only a few transistors called "Small-Scale Integration" (SSI), they used circuits containing transistors numbering in the tens.

SSI circuits were used in early Aerospace project and Missile projects. The two major program of that time, Minuteman missile and Apollo program needed lightweight digital computers for their inertial guidance systems. The Apollo guidance computer motivated the integrated-circuit technology, while the Minuteman missile forced it into mass-production which led SSI to become commercial. These programs acquired almost all of the available integrated circuits from 1960 through 1963, and almost alone provided the demand that funded the production improvements to get the production costs from $1000/circuit (in 1960 dollars) to merely $25/circuit (in 1963 dollars). After the successful implementation in defense industry they began to appear in consumer products a typical application being FM inter-carrier sound processing in television receivers.

In the late 1960s, the next step in the development of integrated circuit was taken with introduction of devices that contained hundreds of transistors on each chip, called Medium-Scale Integration" (MSI).

They were more attractive and fast processing then SSI because, they allowed more complex systems to be produced using smaller circuit boards, less assembly work (because of fewer separate components), and a number of other advantages but they cost little more to produce than SSI devices.

First and second generation microprocessor, computer memories, calculator chips led further development of IC for mass commercial production so Large Scale Integration (LSI) circuits come into picture in the early 1970s that contain tens of thousand transistors on each chip and finally in 1974 to Very Large Scale Integration (VLSI) circuits containing hundreds of thousands transistors on each chip in early 1980s and continues to millions transistors. This led to increase production of chips utilizing then in different new products also the shrinking of chip size and reduction of chip cost.

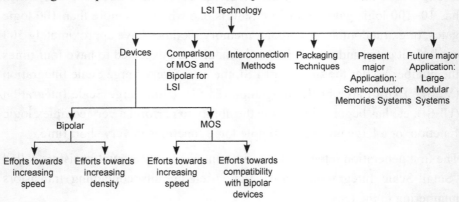

Fig. 1.2 LSI Technology

Figure 1.2 gives the way of dividing discussion on LSI technology on a number of dimensions.

Scale of Integration	No of Transistor	Function	Time
SSI	Fewer than 10	Input and output gates are connected directly to package	1963
MSI	10 to 99	Performs digital function like decoders, adders, registers	1970
LSI	100 to 9999	Includes digital systems such as processors, memory chips and programmable modules	1975
VLSI/ULSI	10000 to 99999	Include large memory array and complex microcomputer chips	1980

In recent years the rate of growth has showed difficulties in defining, designing and processing complicated chips were about 100 million devices /chip available before 2000 and 1 billion in 2011. The devices which are used in today's integrated circuits, primarily CMOS, Bi-CMOS, GaAs and FinFET, in 1980 at the beginning of VLSI era the minimum feature size was 2µm which shrink to 0.1 µm in 2000 and to 0.022µm in 2011. Device miniaturization results in reduced unit cost per function and in improved performance. The device speed has improved by four orders of magnitude since 1960. Higher speed leads to expanded IC functional throughput rates. Digital ICs are able to perform data processing, numerical computation and signal conditioning at 10 and higher gigabit per second rates. Another benefit is the reduction of power consume as the device become smaller so, consume less power and reduces the energy used for each switching operation.

1.3 CLEAN ROOM

In 1965 the chip manufacturing factories were filthy by today's standards and wafer cleaning procedures were unorganized having poorly understood. The chips were manufactured in those days when they were very small, unreliable and contained very few components by today's standards. Defects on a chip tens to reduce yields exponentially as chip size increases, small chip can be manufactured even in quite dirty environment.

The semiconductor devices are fabricated by introducing dopants, often at concentrations of parts per billion and by depositing and patterning thin films on the wafer surface, often with a thickness control of a few nano meter (nm). Such processes are fabricated and reproducible with high accuracy only if stray contaminants can be held to levels below those that affect devices characteristics on chip yield. Modern IC manufacturing units employ clean rooms to control unwanted impurities. Clean room is implemented by building the chips in a clean dust free ambiance having highly filtered air. Apparatus are designed to minimize particle and residual production for that ultra-pure chemicals and highly filtered gases are used.

The numerous developments have been made in shrinking device geometry refer as device miniaturization and in improving manufacturing so that larger chips can be economically built. This development requires that defect control associated within the manufacturing process also improve. The Semiconductor Industry Association (SIA) data has been summarized in Table 1.1.

Table 1.1 Implication of Semiconductor Industry Growth on defect size, density and contamination level

Year of DRAM shipment	1999	2003	2006	2009	2012	2015
Critical defect size	90 nm	65 nm	50 nm	35 nm	25 nm	18 nm
Starting wafer total LLS (cm^{-2})	0.29	0.14	0.06	0.03	0.015	**0.05**
DRAM GOI defect density (cm^{-2})	0.03	0.014	0.006	0.003	0.001	**0.001**
Logic GOI defect density (cm^{-2})	0.15	0.08	0.05	0.04	0.03	**0.01**
Standard Wafer total bulk Fe (cm^{-2})	$1*10^{10}$	Under $1*10^{10}$	Under $1*10^{10}$	Under $1*10^{10}$	Under $1*10^{10}$	Under $1*10^{10}$
Critical metals on wafer surface after cleaning (cm^{-2})	$4*10^9$	$2*10^9$	$1*10^9$	$< 10^9$	$< 10^9$	$< 10^9$
Starting Material Recombination Lifetime (μ sec)	>=325	>=325	>=325	>=450	>=450	>=450

It is obvious that great care must be taken in making sure that the factories in which chips are manufactured are as clean as possible. Even with a ultra clean environment, and even with procedure with clean wafers thoroughly and often, it is not realistic to expect that all impurities can be kept out of silicon wafers. The critical particle size of the impurity/dopant/dust is on the order of half of the minimum feature size of the devices. Particles larger than this size have a high probability of causing a manufacturing defect. There is simply too much processing and handling of the wafers during IC fabrication.

The manufacturing units producing chips must have clean facilities. Particles that might deposit on a silicon wafer and cause a defect may originate from many sources including people touch-dust, machines processing chemicals, process and residual gases. Such particles may be airborne or may be suspended in liquids or gases. It is common to characterize the cleanliness of air in IC facilities by the designation "class 10 or class 100". Figure 1.3 illustrates the meaning of these terms.

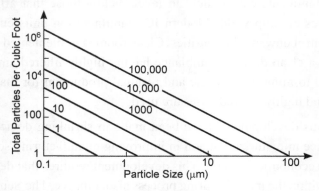

Fig. 1.3 Particle size distribution curve for various classes of clean room. The vertical axis is the total number of particles larger than a given particle size.

1.3 Clean Room

Class 10 classifies that in each cubic foot of air in the manufacturing unit, there are less than 10 total particles having size greater than 0.5 µm. A typical class room of university is about class 100000 while room air in state of the art manufacturing facilities today is typically class 1 in critical areas. This level of cleanliness is obtained through a combination of air filtration and circulation, clean room design and through careful elimination of particular sources.

Particles in the air in a manufacturing plant generally come from several main sources. This includes the people who work in the plant, machines that operate in the plant, and supplies that are brought into the plant. Many studies have been done to identify particle source and the relative importance of various sources. For example people typically emit several hundred particle per minute from each cm^2 of surface area. The actual rate is different for clothing versus skin versus hair but net result is that a typical human emits 5-10 million particles per minute. Most modern IC manufacturing plant makes use of robots for wafer handling in an effort to minimize human handling and therefore particle contamination.

The very first step in reducing particles is to minimize these sources. People in the plant should wear "bunny suits" which cover their bodies and clothing and which lock particle emissions from these sources. Face masks and individual air filters are also worn to prevent exhaling particles into the room air. Few minute air showers at the entrance of the clean room to blow loose particles off from people before they enter as well as clean room protocols are enforced to minimize particle generation. Machines those handles the wafers in the plant are specifically designed to minimize particle generation and materials are chosen for use inside the plant which minimize particle emission.

The source of particles can never be completely eliminated but constant air filtration is used to remove generated particles. This is accomplished by recirculating the air through High Efficiency Particulate Air (HEPA) filters. These filters are composed of thin porous sheets of ultrafine glass fibers (< 0.5 µm diameter). Room air is forced through the filters with a velocity of about 50 cm/sec. large particles are trapped by the filters; small particles impact the fibers as they pass through the filter and stick to these fibers primarily through electrostatic forces. The net results of HEPA filters are 99.98% efficient at removing particles from the air.

Most IC manufacturing facility produces their own clean water on site, starting with water from the local water supply. This water is filtered to remove dissolved particles and organics. Dissolved ionic species are removed by ion exchange or reverse osmosis. The result is high purity (high resistivity in MΩ) water that is used in large quantities in the plant.

Modern Chip manufacturing plant is designed to continuously recirculate the room air through HEPA filters to maintain a class 10 or class 1 ambiance. A typical clean room is shown in figure 1.4.

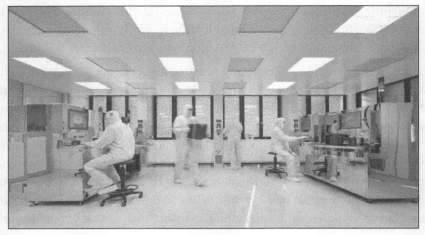

Fig. 1.4 Typical modern cleanroom for IC fabrication.
(Photo courtesy of graphene.manchester.ac.uk)

All mechanical support equipment is located beneath the clean room to minimize contamination from these machines. The HEPA filters are located in the ceiling of clean room. The fans that recirculate the air are normally placed above HEPA filter. Inside cleanroom, finger walls or chases provide a path for air return as well as to bring in electric power, distilled water and gases. The scientist and engineers wear " bunny suits" to minimize particle emission.

1.4 SEMICONDUCTOR MATERIALS

Semiconductors are a class of materials which have the unique properties that their electrical conductivity can be controlled over a very wide range by the introduction of dopants. Dopants are atoms that generally contain either one more or one fewer electrons in their outermost shell than the host semiconductor. They provide one extra electron or one missing electron (a "hole") compared to the host atoms. These excess electron and holes are the carriers which carry current in semiconductor devices. The key to building IC lies in the ability to control the local doping and hence the local electronic properties of semiconductor crystal.

Elemental semiconductors all of which have covalent bonding or sharing arrangement in essence populates the entire outer shell for each atom, resulting in a stable structure in which all electron are bound to atoms at least at very low temperature. This kind of bonding arrangement lies in column IV of

1.4 Semiconductor Materials

periodic table shown in figure 1.5. This same type of bonding arrangement can be produced using mixture of elements from other columns of periodic table known as compound semiconductors. For example GaAs consists of alternating Ga (Column III) and As (Column V) atoms which have an average of four electron per atom, ZnO consist of alternating Zn (Column II) and O (column VI) atoms which have an average of four electron per atom and so the same covalent bonding arrangements works. More complex examples like, $Al_xGa_{1-x}As$, $Hg_xCd_{1-x}Te$, $Al_{1-x}Ga_xAs_yP_{1-y}$ are also possible. Thus nature provides many possible materials which can act as semiconductor.

	III	IV	V	VI
	5 **B** 10.81	6 **C** 12.01	7 **N** 14.01	8 **O** 16.00
	13 **Al** 26.96	14 **Si** 28.09	15 **P** 30.97	16 **S** 32.06
30 **Zn** 65.38	31 **Ga** 69.72	32 **Ge** 72.59	33 **As** 74.92	34 **Se** 78.96
48 **Cd** 112.40	49 **In** 114.80	50 **Sn** 118.70	51 **Sb** 121.80	52 **Te** 127.60

Fig. 1.5 Periodic table portion relevant for elemental and compound Semiconductors

At temperature above absolute zero thermal energy can break some of the covalent bond of semiconductor which creates both a free or mobile electron and a mobile hole. The concentration of electron and holes are exactly equal in pure semiconductor are referred to as intrinsic semiconductor. The conductivity of pure semiconductor depends on the broken covalent bond due to temperature. So the free charge carriers are very few in pure semiconductor. Fortunately semiconductors have the properties that they can be doped with other materials. Doping results in a column V (P, As) or a column III (B, Al) atom replacing a semiconductor atom in the crystal structure. Such dopants either contribute an extra electron (column V) to the crystal, become N-type dopants or they contribute a hole (column III), become p-type dopants. The electron and holes are introduced on a one for one basis by the dopants. Doping could be accomplished by diffusion or ion-implantation, modern IC technology generally uses ion-implantation to dope semiconductor which permits controlled introduction of parts per million to parts per hundred of dopant atoms. As a result conductivity of semiconductor can be controlled over a very wide range, permitting many types of semiconductor devices to be fabricated. The device fabrication involve a large processing steps. Elemental semiconductors (Si, Ge) have advantages over compound semiconductor as they do not decompose on processing. Although Silicon (Si) dominate (about 95%) in semiconductor industry but it is not an optimum choice in every respect. For optoel ectronic purpose Si is not preferred as it is indirect bandgap semiconductor. For this compound semiconductor of direct bandgap is preferred such as GaAs. Which processing technology is most highly developed. There are several reasons for Si to become the preferred elemental semiconductor for IC in the present over Ge. The reasons are listed below as:

1. Si has a larger bandgap (1.1eV) compared to Ge (0.7eV) at room temperature) and because of this the henomenon of thermal electron hole pair generation is smaller in Si then Ge. Which mean that at the same temperature the noise of the Si devices is smaller than the noise of Ge devices.

2. Si is relatively easy and inexpensive to obtain and process so it is cheap whereas Ge is rare material that is typically found with copper, lead or silver deposits so it is expensive and difficult to process comparatively.

3. Unlike Ge, Si forms native oxide (thin layer of SiO_2) on its surface very easily which is a very good insulator and which technologically can be very easily processed. Thin layer of oxide is very useful to form the gates of MOSFET transistors. Ge does not form native oxide layer on its surface so easily and technology to obtain the Ge devices is more complicated.

4. The reverse current in Si flows in order of nano-amperes compared to Ge in which the reverse current in order of micro-amperes because of this the accuracy of non-conduction of Ge diode in reverse bias falls down. The Si diode has large reverse breakdown about 70-100 V compared to Ge which has the reverse breakdown voltage around 50 V.

5. Because of the high band gap temperature stability of Si is good, it can withstand in temperature range typically 140°C to 180°C whereas Ge is much temperature sensitive only upto 70°C.

1.5 CRYSTAL STRUCTURE

The materials used for microelectronics can be divided into three classifications (Amorphous, Single crystal & Poly crystal), depending on the amount of atomic order they possess.

In single crystal materials, almost all of the atoms in the crystal occupy well-defined and regular positions known as lattice sites. In semiconductor products

1.5 Crystal structure

base layer is single crystal provided by the wafer or substrate. If there is an additional single-crystal layer, it is grown epitaxially on substrate.

Amorphous materials are at the opposite extreme the atoms have no long-range order such as SiO_2. Instead, the chemical bonds have a range of lengths and orientations.

The third class of materials is polycrystalline. These materials are a collection of small single crystals randomly oriented with respect to each other. The size and orientation of these crystals often change during processing and sometimes even during circuit operation.

The silicon industry depends on a ready supply of high quality single crystal wafers.

Crystals are described by their most basic structural element a "unit cell" simply arranged in an array, repeated in a very regular manner over three dimensions. The unit cells of interest have cubic symmetry with each edge of the unit cell being the same length.

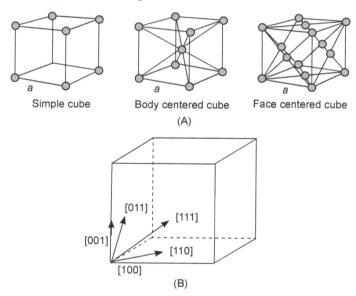

Fig. 1.6 (a) Cubic Crystal lattices b) Crystal orientation in cubic system

Figure 1.6 shows three simple crystal unit cells. All are based on a cubic structure.

Simple Cubic (SC): Polonium crystal exhibits this structure over a narrow range of temperature. The SC cell has atoms at the corners of the cell.

Body Centred Cubic (BCC): Molybdenum, tentalum and tungsten exhibits this crystal. The bcc cell has an additional extra atom in the center of the cube than SC.

Face-Centred Cubic (FCC): This structure is exhibited by a large number of elements such as copper, gold, nickel, platinum and silver. This cell has additional extra atoms in the center of each face of the cube than SC.

The directions in a crystal are identified using a Cartesian coordinate system as $[x,y,z]$. For a cubic crystal, the faces of the cell form planes that are perpendicular to the axes of the coordinate system. The symbol (x,y,z) is used to denote a particular plane that is perpendicular to the vector that points from the origin along the $[x,y,z]$ direction. Figure 1.6 b) shows several common crystal directions. The set of numbers h,k, and l that are used to describe planes in this manner are called the *Miller indices* of a plane. They are found for a given plane by taking the inverse of the points at which the plane in question crosses the three coordinate axes, then multiplying by the smallest possible factor to make h, k, and l integers. The notation $\{h,k,l\}$ is also used to represent crystal planes. This representation is meant to include not only the given plane, but also all equivalent planes. For example, in a crystal with cubic symmetry the (100) plane will have exactly the same properties as the (010) and (001) planes. The only difference is an arbitrary choice of coordinate system and the notation {100} refers to all three.

Silicon and germanium are both group IV elements. They have four valence electrons and need four more to complete their valence shell. In crystals, this is done by forming covalent bonds with four nearest neighbor atoms. None of the basic cubic structures in figure 1.6 would therefore be appropriate. The simple cubic crystal has six nearest neighbors, the body-centered cubic (BCC) has eight, and the face-centered cubic (FCC) has twelve. Instead, group IV semiconductors form in the diamond structure shown in figure 1.7. The unit cell of Si can be constructed by starting with an FCC cell and adding four additional atoms If the length of each side is a, the four additional atoms are located at $(a/4, a/4, a/4)$, $(3a/4, 3a/4, a/4)$, $(3a/4, a/4, 3a/4)$, and $(a/4\ 3a/4, 3a/4)$.

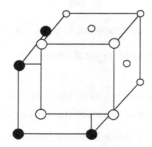

Fig. 1.7 Zincblende Lattice

This crystal structure can also be thought of as two interlocking FCC lattices. Gallium arsenide also forms in this same arrangement; however, when two elements are present, the crystal has a reduced level of symmetry so the structure is then called zincblende.

1.6 CRYSTAL DEFECTS

A perfect crystal that has every atom of the same type in the correct position does not exist. All crystals have some defects which contribute to the mechanical properties of materials. In fact, using the term "defect" is sort of a misnomer since these features are commonly intentionally used to manipulate the mechanical properties of a material as adding alloying elements to a metal is one way of introducing a crystal defect. Nevertheless, the term "defect" will be used to just keep in mind that crystalline defects are not always bad. Semiconductor wafers are highly perfect single crystals. There are several types of defects which are commonly found in crystals. Nevertheless, crystal defects play an important role in semiconductor fabrication. Semiconductor defects or imperfections, can be distributed into four types depending on their dimensionality as describe in figure 1.8. These are

- Point defect
- Line defect
- Area defect
- Volume defect

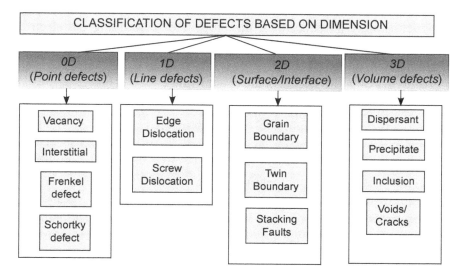

Fig. 1.8 Defects based on Dimensions

Point defects Point defects are simple to visualize and they play crucial roles in impurity diffusion. In layman language anything other than a silicon atom on a lattice site constitute a point defect. By this definition a substitutional doping atom is a point defect and might be referred to as an impurity related defect. Principally point defect can be further divided in two categories: the first is simply missing silicon lattice atom or vacancy and the other is an extra silicon atom.

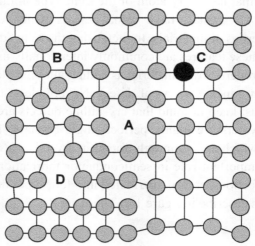

Fig. 1.9 Simple semiconductor defects A-vacancies, B- interstitials, C- substitutional impurities, D- edge dislocation

Point defect include vacancies, interstitials, misplaced atoms, dopant impurity atoms deliberately introduced for the purpose of controlling the electronic properties of the semiconductor and impurity atoms which are used as contaminations during material growth or processing. Figure 1.9 shows different types of point defects in the crystal lattice.

One of the most common types of point defect is a lattice site without an atom. This defect is a vacancy figure 1.9A. Closely related point defect is an atom that resides not on a lattice site, but the spaces between the adjacent lattice positions. It is referred to as an interstitial figure 1.9B. If the interstitial or vacancy atom is of the same material as the atoms in the lattice; it is a self-interstitial. In some cases, the interstitial comes from a nearby vacancy such a vacancy interstitial combination is called a Frenkel defect. The interstitial or vacancy may not remain at the site at which it was created. Both types of defects can move through the crystal particularly under the high temperature ambience that arises during processing condition. Either defect might also migrate to the surface of the wafer where it is annihilated.

1.6 Crystal Defects

The second type of point defect that may exist in a semiconductor is known as extrinsic defect shown in figure 1.9C. This is caused either by an impurity atom at an interstitial site or at a lattice site. In the second case it is referred to as a *substitution impurity*. For example dopant atoms are required to modulate semiconductor conductivity are basically caused by substitutional defects. Substitutional as well as interstitial impurities have a significant impact on device performance. Some impurities that tend to occupy interstitial sites have electronic states near the center of the bandgap. As a result, they are efficient sites for the recombination of electron-hole pair. These recombination centers form depletion region that reduce the gain of bipolar transistors and can cause p–n diodes to leak.

Line Defects: Line defects or dislocations, are lines along which whole rows or columns of atoms in a solid crystal are arranged anomalously. The resulting irregularity in spacing is most severe along a line called the line of dislocation as shown in figure 1.10. Line defects are mostly due to misalignment of ions or presence of vacancies along a line that can be weaken or strengthen solids. When lines of ions are missing in an otherwise perfect array of ions, an edge dislocation appeared which is responsible for the ductility and malleability. In fact movement of edge dislocation often results hammering and stretching of materials. Movements of dislocations give rise to their plastic behavior. Line dislocations usually do not end inside the crystal, they form loops or end at the surface of a single crystal.

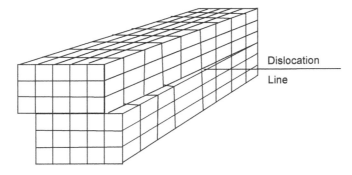

Fig. 1.10 Line dislocation

A dislocation is characterized by the term known as Burgers vector; If you will imagine going around the dislocation line, and exactly going back as many atoms present in each direction as one have gone forward, one will not come back to the same atom where he has started. The Burgers vector points from start atom to the end atom of this journey and this "journey" is known as Burgers circuit in line dislocation theory.

Figure 1.11 shows the Burger circuit journey around an edge dislocation using the sketch of surface of a crystal. A Burger vector is approximately perpendicular to the dislocation line, and the missing line of atoms is somewhere within the block of the Burger journey.

If the misalignment shifts a block of ions gradually downwards or upwards causing the formation of a screw like deformation, a screw dislocation is formed as shown in figure 1.11(b).

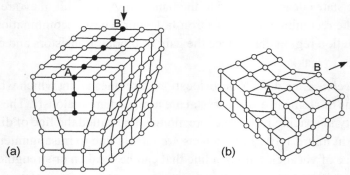

Fig. 1.11 (a) Edge dislocation formation in crystal and (b) screw dislocation formation in crystals

Line defect affects the mechanical properties of the solid in terms of its density as well as deteriorates the structure along a one-dimensional space. Mechanical properties are also affected by the type of line defects. As a result for structural materials, the formation and study of dislocations are particularly important.

Surface Defects: These defects may arise at the boundary between two grains, or merging of two crystals such as small crystal within a larger crystal. The rows of atoms in two different grains may run in slightly different directions, leading to a mismatch across the grain boundary as shown in figure 1.12. The external surface of a crystal also comes under surface defect because the atoms on the surface relocate their positions to accommodate for the absence of neighboring atoms.

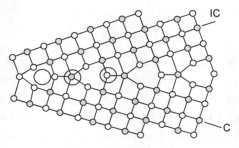

Fig. 1.12 Surface defects

Volume or Bulk Defects

Volume or bulk defects are 3-dimensional defects. These include cracks, pores, external inclusions and several other phases. These defects are generally introduced during fabrication steps. All these defects are capable of acting as stress raisers therefore harmful to parent material's mechanical behavior. However, in some cases foreign particles are added selectively and purposefully to mechanically strengthen the material. The procedure in which foreign particles act as obstacles to movement of dislocations, which facilitates plastic deformation is known as dispersion hardening. The second-phase particles act in two different ways – first particles are either may be cut by the dislocations and secondly the particles resist cutting and dislocations are forced to bypass them. Due to ordered particles strengthening is responsible for the good high-temperature strength on many alloys. However, pores are harmful because they reduce effective load bearing area and act as stress concentration sites.

It is common to divide aggregates of atoms or vacancies into four classes in an imprecise classification that is based on a combination of the size and effect of the particle. The four categories are: (A) **Precipitates**, which area fraction of a micrometer in size and decorate the crystal; (B) **Dispersants** or second phase particles, which vary in size from a fraction of a micrometer to the normal grain size (10-100µm), but are intentionally introduced into the microstructure; (C) **Inclusions**, which vary in size from a few microns to macroscopic dimensions, and are relatively large, undesirable particles that entered the system as dirt or formed by precipitation; (4) **Voids or pores**, which are holes in the solid formed by trapped gases or by the accumulation of vacancies.

Precipitates are small particles that are introduced into the matrix by solid state reactions. The precipitates are used for several purposes, their most useful purpose is to increase the strength of structural alloys by acting as hurdles to the motion of dislocations. Their efficiency in doing this depends on several factors such as their internal properties, their size, and their distribution throughout the lattice. However, their role in the microstructure is to modify the behavior of the crystal matrix rather than to act as separate phases in their own right.

Dispersants are bigger particles that behave as a second phase and influence the behavior of the primary phase. They may be large precipitates, grains, or polygranular particles distributed through the microstructure.

Inclusions are foreign particles or large precipitate particles. They are usually unwanted constituents in the microstructure. Inclusions have a harmful effect on the strength of structural alloys since they are preferential sites for failure.

They are most time harmful in microelectronic devices as they change the geometry of the device by interfering in fabrication, or alter its electrical properties by hosting undesirable properties of their own.

Voids are produced as a result of gases those are trapped during solidification or by vacancy condensation in the solid state. So they are almost always undesirable defects. Their prime effect is to decrease mechanical strength and endorse fracture at small loads.

1.7 SI PROPERTIES & ITS PURIFICATION

More than 90% of earth crust is composed of Silica (SiO_2) or Silicate making Si the second most abundant element on earth after oxygen. It is found in rocks, sand, clays and soils combined with oxygen as SiO_2 or silicates. Si is the fourteen element of the periodic table and is a group IVA element. Pure Si is a dark gray solid with the same crystalline structure as diamond; with each atom covalently bonded to four nearest neighbors. In pure form its lattice constant is 5.43086 Å at 300K. The nearest neighbor distance between Si atom in diamond lattice is 2.35163 Å. Si has a melting point of 2570°F (1410°C), a boiling point of 4271°F (2355°C) and a density of 2.33 g/cm³.

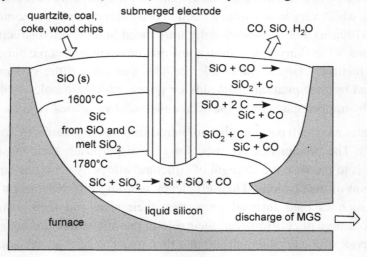

Fig. 1.13 Formation of MGS through arc furnace

The Si industry depends on a ready supply of inexpensive high quality single crystal wafer. IC manufactures typically specify physical parameter (diameter, thickness, flatness, mechanical defect etc.) electrical parameters (N or P type, dopant, resistivity etc.) and finally impurity levels (oxygen and carbon in particular) when purchasing wafers. For single crystalline Si wafer, Si must be refines as well as be converted into crystalline form. This is usually a

1.7 Si properties & its Purification

multistage process beginning with sand (SiO_2) name as Quartzite. Conversion the quartzite to Metallurgical Grade Silicon (MGS) is the initial step in refining process. Figure 1.13 shows the production of MGS from arc furnace. This process usually takes place in a furnace in which the quartzite and carbon source (coal or coke) mixture is heated to temperature approaching 2000°C.

$$2C\,(s) + SiO_2\,(s) = Si\,(l) + 2CO\,(g)$$

The MGS grade Si that results is about 98% pure, with Al and iron being two of the dominant impurities most of MGS is used in manufacturing Al or Silicon Polymers.

To convert the MGS into highly purified polycrystalline Electronic Graded Silicon (EGS) multi stage process is required. Pure EGS generally requires that the doping elements reduced to be in ppm range and carbon impurities should be less than 2 ppm. In MGS major impurities are boron, carbon and residual donors. Seimens Process is one of the major techniques for converting MGS to EGS. The numbers of steps to convert MGS to EGS are: first step MGS reacts with gaseous HCl, commonly by grinding the MGS into a fine powder and then the reaction takes place in the presence of catalyst at elevated temperature. This process can form any number of Si-H-Cl compounds ($SiH_4 \rightarrow$ silane; $SiH_3Cl \rightarrow$ chlorosilane; $SiH_2Cl_2 \rightarrow$ dichlorosilane; $SiHCl_3 \rightarrow$ trichlorosilane or $SiCl_4 \rightarrow$ silicon tetra chloride)

$$Si\,(s) + 3HCl\,(g) \longrightarrow SiHCl_3\,(g) + H_2\,(g) + Heat$$

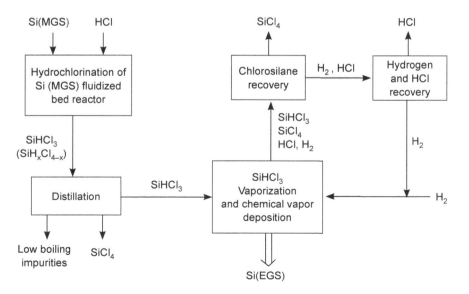

Fig. 1.14 Conversion steps of MGS to EGS

EGS is prepared from the purified $SiHCl_3$ in a Chemical Vapor Deposition (CVD) reactor using a resistance heated rod of Si which serve as the nucleation point called a "Slim rod" for the deposition of polysilicon.

$$2SiHCl_3 \text{ (g)} + 2H_2 \text{ (g)} \longrightarrow 2Si \text{ (s)} + 6HCl \text{ (g)}$$

The polysilicon which is deposited may be several meters long and several hundred mm in diameter.

An alternate process is starting to receive commercial attention for the production of EGS by pyrolysis of silane.

$$SiH_4 \text{ (g)} \xrightarrow{900°C} Si(s) + 2H_2 \text{ (g)}$$

The final material is the EGS that has very small impurity level usually in the order of ppb or 10^{13}-10^{14} cm^{-3} which is close to what is desired for the final single crystal wafer but like MSG, EGS is also polycrystalline in nature. Figure 1.14 shows different steps of conversion of MGS to EGS. Formation of wafers required single crystal Si substrate but EGS is still polycrystalline so it needs to be converted into a single crystal Si ingot. The polysilicon is broken up into pieces to load into crucibles of czochralski crystal growth technique or float-zone crystal growth technique to get Single Crystal Silicon.

1.8 SINGLE CRYSTAL SI MANUFACTURE

A single crystal ingot is used to obtain the final wafers therefore there are mainly two techniques for converting polycrystalline EGS into single crystal Si ingot; these are

1. **Czochralski technique (CZ):** This is the dominant technique for manufacturing large wafers of single crystal Si wafers that are currently used in IC fabrication. Nearly 80-90% Si wafer production manufactures practices CZ technique.
2. **Float zone technique (FZ):** The float zone technique captured 10-20% Si wafer market and is basically used for producing specialty wafers that have low oxygen impurity concentration. This technique is mainly used for small sized wafer production.

1.8.1 Czochralski Crystal Growth Technique

CZ technique is invented by polish scientist J. Czochralski in 1918. A Czochralski crystal growth apparatus also called a puller as shown in figure 1.15. The pullar has four subsystems as follows.

1. **Furnace:** Crucible, susceptor and rotation mechanism, heating element, power supply, and chamber.

1.8 Single crystal Si manufacture

2. Crystal-pulling mechanism: Seed shaft or chain, rotation mechanism and seed chuck
3. Ambient control: Gas source, purge tube, flow control and exhaust system
4. Control system: Sensors microprocessor and outputs

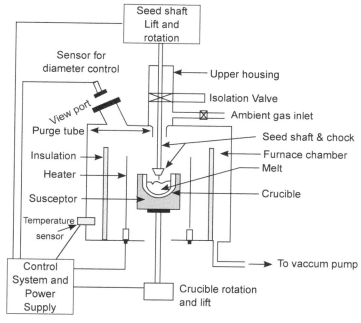

Fig. 1.15 Czochralski crystal growth system

Furnace: The most important component of the growing system is the crucible. The crucible material should have thermal hardness, chemically stability, high melting point since it contain the melt of Si and it should be unreactive with molten Si for reusable purpose. Commonly used high temperature material such as TiC, TaC, SiC are unsuitable because of introducing unacceptable levels of impurities into the crystal. The remaining choices for crucible are Si_3N_4 and fused SiO_2.

Fused silica (SiO_2) reacts with Si releasing Si and O_2 into the melt. The dissolution rate is quite substantial being in the range $(8-25)*10^7$ g/cm².s. The actual rate of erosion is a function of the convection conditions in the melt and temperature. Large portion of the oxygen in the melt outflows by the formation of gaseous Silicon Monoxide (SiO) which on condensing creating the cleanliness problem in the puller. The purity of silicon is also affected by the silica purity because SiO_2 can captivate sufficient acceptor impurities to limit the maximum values of resistivity of the Si that is being grown. The presence of carbon in the melt also accelerates the dissolution rate up to twofold.

$$C\,(s) + SiO_2\,(s) \longrightarrow SiO\,(g) + CO\,(g)$$

Crucible for large CZ pullers have a diameter-to-height ratio of approximately one or slightly greater: common diameters are 25, 30, 35 and 45 cm for charge size of 12, 20, 30 and 45 kg respectively. Wall thickness of 0.25 cm is used but the silica is sufficiently soft to require the use of a susceptor for mechanical support. Upon cooling, the thermal mismatch between residual Si and SiO_2 usually results in the fracture of crucible. The feasibility of using Si_3N_4 as a crucible material has been demonstrated using CVD-deposited nitride. It is attractive as a means of eliminating oxygen from crucible-grown crystal. However even the nitride is eroded resulting in a doping of crystal with nitrogen, a weak donor. CVD nitride is the only form of nitride with sufficient purity for crucible use.

The main functions of susceptor as mentioned previously, are to support the silica crucible and provide better thermal conditions. A high purity nuclear graded graphiteis the preferred choice material for susceptor. Preventing contamination of the crystal from impurities that would be volatilized from graphite at temperature involved is necessity for high purity. The position of susceptor is on a pedestal whose shaft is connected to a motor that provides rotation. The whole assembly can usually be raised up and down to keep the melt level equidistant from a fixed reference point which is required for automatic diameter control.

The chamber covering the furnace must meet conditions that it should provide easy access to the furnace components to facilitate maintenance and cleaning. The furnace must be designed in such a manner to prevent contamination from the atmospheric ambience as well as vapor pressure generated due to heating will not be a factor for crystal. As a rule, the hottest parts of the puller are water cooled. Insulation is usually provided between the heater and chamber wall.

The melt the charge, radio frequency (induction heating) or resistance heating has been used. Induction heating is useful for small melt sizes, but resistance is used exclusively in large pullers. Resistance heaters at power levels involved are generally smaller, cheaper, easier to instrument and more efficient. A graphite heater is connected to a DC power supply.

Crystal pulling mechanism

Crystal pulling mechanism must provide minimal vibration with great precision. The control on growth process must have two parameters: A) pull rate B) crystal rotation; the lead screws are often used to withdraw and rotate crystal. This method provide centers the crystal relative to the crucible, but may require an excessively tall apparatus if the grower is to produce long crystal. Since precise mechanical tolerance is difficult to maintain over a long shaft, pulling with a cable may be necessary. Using cable makes difficult of centering crucible & crystal but provides a smooth pulling action. The crystal leaves the furnace through a tube where the crystal is cooled by passing ambient gas in direction along the surface the name of the tube is purge tube. After purge tube the crystal enters an upper chamber that is separated from the furnace by an isolation valve.

Ambient control

Inert gas or vacuum ambience should be provided for Czochralski growth of Si because 1) to prevent erosion, hot graphite part must be protected from oxygen 2) molten Si should not react with the gas around. Growth in a vacuum meet these requirement; it also has the advantage of removing SiO from the system, thus preventing its buildup inside the furnace chamber. Si crystal growth in a gaseous atmosphere commonly uses inert gases such as He and Ar but Ar is preferred in industrial scale.

Control system

The control system can take many forms to control process parameters like pull rate, rotation speed crystal diameter and temperature. The large thermal mass of the melt generally precludes any short-term control of the process based on temperature. To control the diameter an infrared temperature sensor can be focused on the melt-crystal interface and used to detect changes in the meniscus temperature. The sensor output is linked to the pulling mechanism and controls the diameter by varying the pull rate. The trend in control systems is to use digital microprocessor based system.

Crystal growing theory

Growing crystals involves conversion from solid, liquid or gas phase into crystalline solid phase. The growth of CZ crystal involves conversion of liquid to solid phase known as solidification of atoms at an interface. Figure 1.16 shows the transport process of atoms and temperature gradient involves. Macroscopically, the model for heat transfer conditions about the interface can represented by following equation

$$L\frac{dm}{dt} + k_l \frac{dT}{dx_1} A_1 = k_s \frac{dT}{dx_1} A_2 \qquad \ldots (1.1)$$

Where L is latent heat of fusion, $\dfrac{dm}{dt}$ is the mass solidification rate, T is temperature, K_l & K_s are thermal conductivities of liquid and solid respectively, $\dfrac{dT}{dx_1}$ and $\dfrac{dT}{dx_2}$ are thermal gradients and A_1 and A_2 are the areas of the isotherms at positions 1 & 2 respectively.

From equation (1) the maximum pull rate of a crystal can be deduced by the condition of zero thermal gradient in the melt (i.e. $\dfrac{dT}{dx_1}$). Converting the mass solidification rate to a growth rate using density and area yields

$$V_{max} = \dfrac{k_s}{Ld}\dfrac{dT}{dx} \qquad \ldots (1.2)$$

Where V_{max} is the maximum pull rate and d is the density of solid Si.

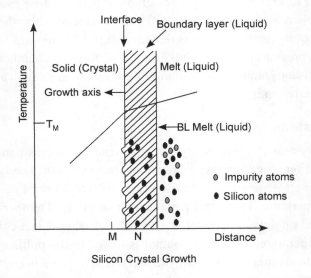

Fig. 1.16 Transport process, solidification and Temperature Gradient involved in Czochralski growth process.

In reality, the maximum pull rate is not commonly used. The pull rate influences the incorporation of impurities into the crystal and is a factor in defect generation which affects the crystalline quality. The material near the melt has a very high density of point defects. Quickly cooling of solid is desirable to prevent these defects from agglomerating but such rapid cooling lead to large thermal gradients on surface (and therefore large stresses) will occur in the crystal, particularly for large-diameter wafers. When the temperature gradient is small in the melt the heat transferred is the latent heat of fusion. Which implies that crystal diameter varies inversely to the pull rate. The pull

1.8 Single crystal Si manufacture

rates acquire in practice should be 30 to 50% slower than maximum values suggested by thermal consideration.

The growth rate of the crystal is perhaps the most important growth parameter but actually distinct from the pull rate as pull rate is the indication of net solidification rate whereas growth rate is instantaneous solidification rate. The two differ because of temperature fluctuations near the interface. The growth rate influence dopant distribution and defect structure in the crystal on microscopic scale. Pull rate affects the defect properties of CZ crystals in the following way: the condensation of thermal point defects into small dislocation loops occurs as the crystal cool from the solidification temperature above 950°C. A pull rate of 2 mm/min eliminate defect formation by quenching the point defect in the lattice before they can agglomerate for diameter above 75 mm. Pull rate is also a factor of determining the growing interface shape as are the melt radial temperature gradient and crystal surface cooling conditions.

Both types of Impurities, unintentional and intentional, are produced into the silicon ingot. Intentional dopants are originate from the melt during crystal growth, while unintentional impurities got introduced from the crucible, ambient, etc. In the melt and solid, common impurities have different solubilities. The equilibrium segregation coefficient k_o is defined as the ratio of the equilibrium concentration of the impurity in the solid to that in the liquid,

i.e.
$$k_0 = \frac{C_s}{C_l} \quad \ldots (1.3)$$

The equilibrium segregation coefficients for various common dopants and impurities are listed in table 1.2. As the crystal is being pulled, all equilibrium segregation constant are below unity (less than one), implies that the impurities preferentially isolate to the melt and the melt becomes progressively enriched with these impurities.

Table 1.2 Segregation coefficients for usual impurities in silicon

Impurity	Al	As	B	C	Cu	Fe	O	P	Sb
k_o	0.002	0.3	0.8	0.07	$4*10^{-6}$	$8*10^{-6}$	0.25	0.35	0.023

The distribution of an impurity in the grown crystal can be described mathematically by the normal freezing relation:

$$C_s = k_0 C_0 (1-X)^{(k_0 - 1)} \quad \ldots (1.4)$$

where X is the fraction of the melt solidified,

C_o = initial melt concentration,

C_s = solid concentration,

1.8.2 Float zone technique (FZ Technique)

For small wafer production having low oxygen impurity, the FZ or float zone technique is most widely used. The principal difference between FZ and CZ process is that no crucible is used. This markedly reduces impurity levels in the resulting crystal, particularly oxygen and makes it easier to grow high resistivity material. Carrier concentrations as low as 10^{11} cm^{-3} have been achieved with the float zone method. The basic feature of this growth technique is that the molten part of the sample is supported entirely by the solid part so no need for a crucible. The process is shown in figure 1.17.

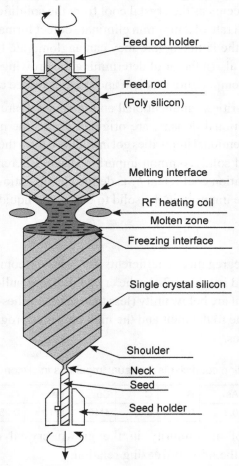

Fig. 1.17 Schematic diagram of Float zone process

In the FZ process, A EGS rod which is polycrystalline in nature clamped at both ends with bottom end fused with the seed of desired single crystal orientation. This is taken in an inert gas furnace and then melted along the length of the rod by a traveling radio frequency (RF) coil. RF coil provides power which generates large current in the Si and locally melts it through I^2R heating. Usually the molten zone is about 2 cm long. RF field generated levitation and surface tension keep the system stable. If the seed end zone is initiated to melt and the rod is slowly moved up then solidifying region has the same orientation as of the seed. For reduction of gaseous impurities, the furnace is filled with an inert gas like argon. Also, since the process requires no crucible so it can be used to produce oxygen free Si wafers. The difficulty is to extend this technique for large wafers, since the process produces large number of dislocations. The process produces large number of dislocations so it is used for small specialty applications requiring low oxygen content wafers. Doping of the crystal can be accomplished either by starting with doped poly-silicon rod, a doped rod, or by maintaining a gaseous ambient during the FZ process that contains a dilute concentration of desired dopant. A disadvantage of float zone growth is the struggle of introducing identical concentration of dopants. There are four methods that can be used: core doping, pill doping, gas doping, and neutron transmutation. The starting material of Core doping is a doped Poly-silicon rod. On top of this rod, additional undoped poly-silicon is deposited until the average desired concentration is reached. The process can be repeated through several generations if necessary. Core doping is the preferred process for boron because its diffusivity is high and because it does not tend to evaporate from the surface of the rod. The concentration of boron in a boule is quite uniform after neglecting the first few melt lengths. Doping is accomplished through the use of gases doping material such as PH_3, $AsCl_3$, or BCl_3. The gas may be injected as the poly-silicon rod is deposited, or it may be injected at the molten ring during the float zone refining. Pill doping is provided by drilling a small hole in the top of the rod and inserting the dopant in the hole. If the dopant has a small segregation coefficient then it will be carried with the melt and passes throughout the length of the boule resulting in modest non-uniformity. Gallium and indium doping work well in this manner. Finally, for light n-type doping, float zone silicon can be doped through a process known as transmutation doping. In this process, the boule is exposed to a high brightness neutron source.

1.9 SILICON SHAPING

Silicon is a hard, brittle material. Industrial-grade diamond is the most suitable material for shaping and cutting silicon, although SiC and Al_2O_3 have also been used and even materials like SiO_2 are used. Commercial wafers or ready to use wafer processing firstly requires six machining operations, two chemical operations and one or two polishing operations to convert silicon ingots into polished wafers. A finished wafer is subject to a number of dimensional tolerance, dictated by the needs of the device fabrication technology. The motivation for standards is two fold (*i*) they help to standardize wafer production, resulting in efficiency and cost saving (*ii*) produces of process equipment and fixtures benefit from knowing the wafer dimensions when designing equipment.

The process of creating the individual wafers beings with shaping the grown crystal or boule to a uniform diameter has been describe by the process flow shown in figure 1.18. Modern crystal growers can not maintain perfect control over the crystal diameter during growth so the crystal is normally grown slightly oversized and trimmed to desired final diameter.

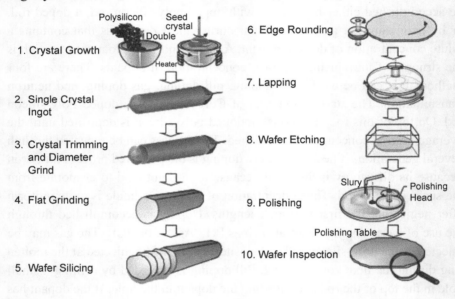

Fig. 1.18 Process flowing steps of shaping of ingot to wafer

Ingot Trimming and Slicing

After getting Si crystal ingot from CZ or FZ processes the next step is shaping operations of ingot which consist of two steps

1. The front and back ends (i.e. seed and tang ends) of the ingot are removed
2. The surface of the ingot is earthed to obtain uniform and constant diameter across the length of the ingot

Before further processing, the ingots have to pass through resistivity and orientation check. Resistivity is checked to confirm the dopant concentration

1.9 Silicon Shaping

along the length of the ingot to ensure uniformity by a four point probe technique discussed in Chapter 6 of Diffusion. Those portion of the ingot who fail the resistivity and perfection evaluations are cut away. The diameter is commonly 100, 150, 200 or 300 mm. Orientation of the ingot is measured by X-Ray Direction (XRD) method at the ends to know the type of wafer and shaping them by flat grinding. A crystallographic orientation flat is also ground along the length of the ingot and defined by two types of flats:

1. **Primary flat**-defines specific crystal direction and act as a visual reference to the orientation of the wafer
2. **Secondary flat**- defines for identification of the wafer, dopant type and orientation

On viewing the wafers based on flats one can easily analysis the type of wafer as described in figure 1.19.

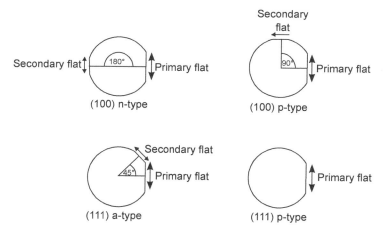

Fig. 1.19 Identify the wafers based on the flats

After grinding process the ingot is then sliced using a large-diameter stainless steel saw blade with industrial diamonds embedded into the inner-diameter cutting edge. This will help in producing circular slices or wafers that are about 600 to 1000 µm thick but the surfaces become quite rough so it has to be lapped to produce a flat surface. Slicing is important because it determines four wafer parameters: surface orientation, thickness for mechanical stability, taper and bow. Wafers of <100> orientation are normally cut "on orientation". The tolerances allowed for orientation do not adversely affect MOS device characteristics such as interface trap density. The primary flat is by convention, oriented perpendicular to <110> direction. In manufacturing, the IC that are later built on these wafers are normally rectangular and are lined up parallel or perpendicular to the (110) flats on the wafer. This means that the edges of the chips are on {110} crystal planes. When the completed chips are ready to be separated for packaging a dicing operation is performed in which the scribe lines between adjacent chips are partially sawn through the individual chips can then be separated by simply breaking the wafer mechanically along these scribes lines. Si actually cleaves naturally along {111} planes. In (100) crystal, the {111} planes meet the surface at an angle of 54.7° along the <110> directions which are at right angles to each other. Thus placing the scribe lines parallel and perpendicular to the wafer flat results in easy cleaving. The two common surface orientations are usually produced using seeds of the appropriate orientation, which then allows cutting of the individual wafers perpendicular to boule wafers of <100> orientation are usually cut "on orientation". The tolerances allowed for orientation do not adversely affect MOS devices characteristics such as interface trap density. The other common orientation <111> is usually cut "off orientation" (by about 3°) as required for epitaxial processing. The wafer thickness is essentially fixed by slicing although the final value depends on subsequent shaping operation. The thickness of the water required to provide adquate mechanical support during IC manufacture 200 mm diamater waters usually about 725 μm thickness then final finished from they must be sawed somewhat thicker than this, however in order to allow for losses that will occur during subsequent lapping, etching and polishing typically sawn wafer thickness would be about 850 μm. The saw blade itself about 400 μm thick and so this thickness is lost from the boule as Si dust every time a wafer is cut. Thus including losses at the seed and tail end of the crystal only about 50% of the boule ends up in wafer form. A major concern is slicing is the blade's continued ability to cut wafers from the crystal in very flat planes. If the blade deflects during slicing this will not be achieved. So positioning a capacitive sensing device near the blade, the blade position and vibration in the blade can be monitored and higher quality cutting achieved. A mechanical two-sided lapping operation, performed under pressure using a wafer with flatness uniform to within 2 μm. A final shaping step is edge controlling, where a radius is ground on the rim of the wafer. The process is usually done in casselte-fed high speed equipment. During device fabrication, edge rounded wafers develop fewer edge chips and aid in controlling the photoresist at the wafer edge. Chipped edges act as places where dislocations can be introduced during thermal cycles and as places where wafer fracture can be initiated. The silicon particles starting from the chipped edge if present on the wafer surface, can add to the defect density of the Integrated circuit process and thus reducing yield.

Etching

The shaping operations leave the surface and edges of the wafer damaged and contaminated with the depth of work damage depending on the specifics of the machining operations. The damaged and contaminated regions are on the order of 10 µm deep can be removed by chemical etching. In past mixtures of three acids hydrofluoric, nitric and acetic acids have been used but utilization of alkaline etching using potassium or sodium hydroxide also in most widely use.

The process equipment contains an acid sink, which uses a tank to keep the etching solution and two or more positions for rinsing the wafers through water. To maintain uniformity, wafer is rotated during acid etching using best available process equipment. Processing is usually performed with a substantial overtake to assure all damage is removed. Removing 20 µm per side is typical. The etching process is checked frequently by gauging wafers for thickness before and after etching.

The etching process involves oxidation-reduction step followed by dissolution of an oxidation product. In the hydrofluoric, nitric, and acetic acid etching system nitric acid is the oxidant and hydrofluoric acid dissolves the oxidized products according to the following reactions:

$$3Si\,(s) + 4HNO_3\,(l) + 18HF\,(l) \longrightarrow 3H_2SiF_6\,(l) + NO\,(g) + 8H_2O\,(l)$$

In HF rich solutions the reaction is made limited at the oxidation step. This oxidation reaction is very much sensitive to doping orientation and defect structure of the crystal. The use of HNO_3 rich mixture provides rate limiting etching and preferred for removing work damage. A mixture of HNO_3 (79% wt.), HF (49% wt.) and CH_3COOH acids with ratio in 4:1:3 is a common etch. With wafer size of larger diameter dimensional uniformity are introduced by lapping, is not maintained to be compatible with surface flatness in polishing. The hydrodynamics of rotating a large size wafer in solution do not allow for a uniform boundary layer, this results in induction of a taper is into the wafer. The projection lithography places demands on surface flatness that makes the use of alkaline etching compulsory. In alkaline etching is apparently dominated by the surface orientation. As in acid etching the reaction is twofold when a mixture of KOH/H_2O or $NaOH/H_2O$ is used. A typical formulation uses KOH and H_2O is a 45% wt. solution (i.e. 45% KOH and 55% H_2O) at 900°C to achieve an etch rate of 25 µm/min for {100} surfaces.

Polishing

The wafer then undergoes a mirror like finish by making a coat of polish on it. The process requires considerable operator attention for loading and unloading. It can be conducted as a single wafer or batch wafer process depending on the equipment. The process involves a processing pad made of artificial fabric such as polyester felt, polyurethane laminate. Wafers are mounted on a fixture, pressed against the pad under high pressure and rotated relative to the pad. This polishing is achieved by using Al abrasive powders of small gravel size to a final 1 μm diameter or a mixture of polishing slurry and water, dripped onto pad. Slurry is a colloidal suspension of fine SiO_2 particles in an aqueous solution of sodium hydroxide. After applying the polish, the wafer remained with a surface damage of around 2 μm deep. This damage is removed by applying an additional chemical etching stage. This additional stage sometimes may be treated with the final polishing stage.

The other side of the wafer is applied a normal lapping procedure to provide a some percentage of flat surface with agreeable parallelism. Once the wafer polishing operation is complete, the next step involves wafers cleaning the wafer, then drying, and after that wafer becomes now ready to be used for the next processing steps.

1.10 WAFER PROCESSING CONSIDERATIONS

Once the wafer is ready after polishing, following sequence of step is applied for next step.

Chemical Cleaning: After polishing is complete, the wafers are cleaned thoroughly. This cleaning operation is performed to get rid of any heavy metals or organic films. The widely used cleaning agents for this purpose are mixtures of $NH_4OH - H_2O_2$, $HCl - H_2O_2$, and, $H_2SO_4 - H_2O_2$. Although all of the above described solutions are good to remove metallic impurities, but, out of these, the HCl, H_2O_2, mixture is reported best in literature.

Gettering Treatment: The metallic impurities in transition group elements are located at the interstitial or substitutional lattice sites and act as generation-recombination centers. These silicid are found to be electrically conductive. The precipitated forms of these defects and impurities are generally silicide. While referring VLSI or ULSI circuits, the performance of these transition group elements decreases, especially in the case of DRAM and narrow-base BJTs, as both are much sensitive to conductive impurity precipitates. To remove the impurities generally a process known as gettering treatment is carried out. By using gettering process wafer can be made free from impurities or defects from the regions where devices are fabricated. When the wafers are developed with sinks for next level of device processing, the defects and impurities are removed with the help of re-gettering process.

1.11 SUMMARY

Integrated circuits have developed with incredible levels of complexity, exceeding 11 billion transistors per chip. This introductory chapter has presented a historic review of Integrated Circuits from the first generation SSI to VLSI. In the VLSI device arena this chapter reviewed some of the most basic properties of semiconductor materials. An introduction to phase diagrams was given for the crystal and a basic description of point, line, volume and area defects of the crystal was also given because these parameters influence the electrical properties and mechanical properties of the semiconductor. In the second half of the chapter, crystal growth methods for silicon were presented. Czochralski growth is the most common method for preparing silicon crystals. Another growth technique for silicon is the float-zone process which offers lower contamination than that normally obtained from the czochralski technique. Float-zone crystals are used mainly for high-power, high-voltage devices where high resistivity materials are required. Flote-zone technique produces mainly small diameter wafers (>150 mm) while czochralski produces large diameter wafer such as 350 mm. After a crystal is grown, it usually goes through wafer-shaping operations to give an end product of highly polished wafers with specified diameter, thickness and surface orientation.

PROBLEMS

1. Calculate the number of gallons of HF and HNO_3 acid needed to remove the work damage from 4000 wafers of 150 nm diameter.
2. Calculate the boron concentration in the crystal that would lead to misfit dislocation formation at a temperature of 1100°C.
3. A CZ melt is simultaneously doped with boron to a level of 10^{16} atoms/cm^3 and phosphorous to a level of 9×10^{16} atoms/cm^3. Does a pn junction form during growth? If so, at what fraction solidified?
4. What is clean room? Why it is required? What are international standards of a typical clean room.
5. List various types of crystal defects that found in a lattice.
6. What is gettering process?

REFERENCES

1. Digest of the IEEE International Solid-State Circuits Conferences, held in February of each year. (http://www.sscs.org/isscc)

2. C.L. Yaws, R. Lutwack, L. Dickens, and G. Hsiu, "Semiconductor Industry Silicon: Physical and Thermodynamic Properties," *Solid State Technical.*,24, 87 (1981).
3. J.C. Brice, "Crystal Growth Processes", Wiley, New York, (1986).
4. W.R. Runyan, "Silicon Semiconductor Technology", McGraw-hill, New York, (1965).
5. R.B. Hering, "Silicon Wafer Technology - State of the Art 1976," *Solid State Technical*, 19, 37 (1976).
6. P.F. Kane and G.B. Larrabee, "Characterization of Solid Surfaces", Plenum Press, (1974).
7. D.K. Schroder, "Semiconductor Material and Device characterization", John Wiley & Sons, (1990).
8. The international Technology Roadmap for Semiconductors, The Semiconductor Industry Association (SIA), San Jose, CA, (1999).
9. M. Stavola, J. R. Patel, L. C. Kimerling, and P. E. Freeland, "Diffusivity of Oxygen in Silicon at the Donor Formation Temperature," *Appl. Phys. Lett.* 42:73 (1983).
10. W.J. Taylor, T.Y. Tan, and U. Gosele, "Carbon Precipitation in Silicon: Why Is It So Difficult?" *Appl. Phys. Lett.* 62:3336 (1993).
11. T. Fukuda, "Mechanical stength of Czochralski silicon crystals with carbon concentrations from 10^{14} to 10^{16} cm^{-3},"*Appl. Phys. Lett.* 65:1376 (1994).
12. X. Yu, D. Yang, X. Ma, J. Yang, Y. Li, and D. Que, "Grown-in Defects in Nitrogen doped Czochralski Silicon," *J. Appl. Phy.* 92:188 (2002).
13. K. Sumino, I. Yonenaga, and M. Imai, "Effects of Nitrogen on Dislocation Behavior and Mechanical Strength in Silicon Crystals," *Appl. Phys. Lett.* 59:5016 (1983).
14. D. Li, D. Yang, and D. Que, "Effects of Nitrogen on Dislocations in Silicon During Heat Treatment," *Physica B* 273-74:553 (1999).
15. D. Tian, D. Yang, X. Ma, L. Li, and D. Que, "Crystal Growth and Oxygen Precipitation Behavior of 300 mm Nitrogen-doped Czochralski Silicon," *J. Cryst. Growth,* 292:257 (2006).
16. G.K. Teal, "Single Crystals of Germanium and Silicon—Basic to the Transistor and the Integrated Circuit," *IEEE Trans. Electron Dev.*, 23:621 (1976).
17. W. Zuhlehner and D. Huber, "Czochralski Grown Silicon,' *Crystals 8,* Springer-Verlag, Berlin, (1982).
18. S. Wolf and R. Tauber, "*Silicon Processing for the VLSI Era,*" Vol. 1, Lattice Press, Sunset Beach, CA, (1986).
19. S.N. Rea, "Czochralski Silicon Pull Rates," *J. Cryst. Growth* 54:267 (1981).
20. W.C. Dash, "Evidence of Dislocation Jogs in Deformed Silicon," *J. Appl. Phys.* 29:705 (1958).

21. W.C. Dash, "Silicon Crystals Free of Dislocations," *J. Appl. Phys.* 29:736 (1958).
22. W.C. Dash, "Growth of Silicon Crystals Free from Dislocations," *J. Appl. Phys.* 30:459 (1959).
23. T. Abe, N.G. Einspruch and H. Huff, "Crystal Fabrication," in *VLSI Electron— Microstructure Sci.* 12, Academic Press, Orlando, F2, (1985).
24. W. Von Ammon, "Dependence of Bulk Defects on the Axial Temperature Gradient of Silicon Crystals During Czochralski Growth," *J. Cryst. Growth* 151:273 (1995).
25. K.M. Kim and E.W. Langlois, "Computer Simulation of Oxygen Separation in CZ/MCZ Silicon Crystals and Comparison with Experimental Results," *J. Electrochem. Soc.* 138:1851 (1991).
26. K. Hoshi, T. Suzuki, Y. Okubo, and N. Isawa, "Extended Abstracts of E.C.S. Spring. Meeting," *Electrochem. Soc. Ext. Abstr.* St. Louis Meet., 811 (1980).
27. J.B. Mullin, B.W. Straughan, and W.S. Brickell, "Liquid encapsulation crystal pulling at high pressures," *J. Phys. Chem. Solid*, 26:782 (1965).
28. I.M. Grant, D. Rumsby, R.M. Ware, M.R. Brozea, and B. Tuck, "Etch Pit Density, Resistivity and Chromium Distribution in Chromium Doped LEC GaAs," *Semi-Insulating III-V Materials*, Shiva Publishing, Nantwick, U.K., 98 (1984).
29. K.W. Kelly, S. Motakes, and K. Koai, "Model-Based Control of Thermal Stresses During LEC Growth of GaAs. II: Crystal Growth Experiments," *J. Cryst. Growth* 113(1-2):265 (1991).
30. R.M. Ware, W. Higgins, K.O. O'Hearn, and M. Tiernan, "Growth and Properties of very Large Crystals of Semi-Insulating Gallium Arsenide," *GaAs IC Symp.*, 2:54 (1996).
31. P. Rudolph and M. Jurisch, "Bulk Growth of GaAs: An Overview," *J. Cryst. Growth* 198-199:325 (1999).
32. S. Miyazawa, and F. Hyuga, "Proximity Effects of Dislocations on GaAs MESFET," *IEEE Trans. Electron. Dev.*, 3:227 (1986).
33. R. Rumsby, R.M. Ware, B. Smith, M. Tyjberg, M. R. Brozel, and E. J. Foulkes, "Technical Digest of 1983 GaAs IC Symposium", Phoenix, 34 (1983).
34. H. Ehrenreich and J. P. Hirth, "Mechanism for Dislocation Density Reduction in GaAs Crystals by Indium Addition," *Appl. Phys. Lett.* 46:668 (1985).
35. G. Jacob, *"Proc. Semi-Insulating III-V Materials,"* Shiva Publishing, Nantwick, U.K., 2 (1982).
36. C. Miner, J. Zorzi, S. Campbell, M. Young, K. Ozard, and K. Borg, "The Relationship Between the Resistivity of Semi-Insulating GaAs and MESFET Properties," *Mat. Sci. Eng B.*, 44:188 (1997).

37. K. Hoshi, N. Isawa, T. Suzuki, and Y. Okubo, "Czochralski Silicon Crystals Grown in a Transverse Magnetic Field," *J. Electrochem. Soc.,* 132:693 (1985).
38. T. Suzuki, N. Izawa, Y. Okubo, and K. Hoshi, *"Semiconductor Silicon 1981,"* 90 (1981).
39. R.N. Thomas, H.M. Hobgood, P.S. Ravishankar, and T.T. Braggins, "Melt Growth of Large Diameter Semiconductors: Part I," *Solid State Technol.* 33:163 (April 1990).
40. S. Sze, *"VLSI Technology,"* McGraw-Hill, New York, (1988).
41. N. Kobayashi, "Convection in Melt Growth—Theory and Experiments," *Proc. 84th Meet. Cryst. Eng.* Jpn. Soc. Appl. Phys., 1 (1984).
42. M. Itsumi, H. Akiya, and T. Ueki, "The Composition of Octahedron Structures Thtat Act as an Origin of Defects in Thermal SiO_2 on Czochralski Silicon," *J. Appl. Phys.* 78:5984 (1995).
43. H. Ozoe, JS. Szymd and K. Suzuki, "Effect of a Magnetic Field in Czochralski Silicon Crystal Growth," in *Modelling of Transport Phenomena in Crystal Growth,"* MIT Press, Cambridge, MA, (2000).
44. R.E. Kremer, D. Francomano, G.H. Beckhart, K.M. Burke, and T. Miller, *Mater. Res. Soc. Symp. Proc. 144:15 (1989).*
45. C.E. Chang, V.F.S. Kip, and W.R. Wilcox, "Vertical Gradient Freeze Growth of GaAs and Naphthalene: Theory and Practice," *J. Cryst. Growth* 22:247 (1974).
46. R.E. Kremer, D. Francomano, B. Freidenreich, H. Marshall, K.M. Burke, A.G. Milnes and C. J. Miner, *"Semi-Insulating Materials 1990,"* Adam-Hilger, London, (1990).
47. W. Gault, E. Monberg, and J. Clemans, "A Novel Application of the Vertical Gradient Freeze Method to the Growth of High Quality III-V Crystals," *J. Cryst. Growth,* 74:491 (1986).
48. E. Buhrig, C. Frank, C. Hannis, and B. Hoffmann, "Growth and Properties of Semi-Insulating VGF-GaAs,"*Mat Sci. Eng. B.* 44:248 (1997).
49. R. Nakai, Y. Hagi, S. Kawarabayashi, H. Migajima, N. Toyoda, M. Kiyama, S. Sawada, N. Kuwata, and S. Nakajima, "Manufacturing Large Diameter GaAs Substrates for Epitaxial Devices by VB Method," *GaAs IC Symp.,* 243 (1998).
50. W. Keller and A. Muhlbauer, *"Float-Zone Silicon,"* Dekker, New York, (1981).

2
Epitaxy

2.1 INTRODUCTION

VLSI application requires that fabricated IC should have minimum latch-up when they are powered, for that it require lightly doped thin film single crystal silicon form on top of the heavily doped single crystal silicon. When a lightly doped crystalline layer is grown over a heavily doped substrate by keeping collector resistance low, a higher breakdown voltage can be achieved as well as a higher operating speed and improved bipolar performance. In today's world silicon epitaxy become necessary to produce junctions (P-N Junctions), devices (Diodes, BJTs, CMOS, BiCMOS and FinFET) as well as compound semiconductors because it is affordable to create high quality crystal growth of technologically important materials. Epitaxy helps to minimize the occurrence of latch-up; improve performance of devices and better control of the concentrations of doping on the devices can also be gained.

The term epitaxy refers to the "Growth of a crystalline layer on (epi) the surface of a crystalline substrate and crystallographic orientation of the substrate surface imposes a crystalline order (taxis) onto the growing film" i.e. the grown film have a crystal structure with certain thickness. The grown film crystal structure may be differs from their bulk so epitaxial deposition has facility to add and arrange atoms upon the crystal surface.

Epitaxy is the regularly oriented controlled growth of one crystalline material upon another. So substrate material act as a seed and the process take place for below the melting temperature. The commercial importance of epitaxy comes mostly from its use in the growth of semiconductor materials for forming layers and quantum wells in electronic and photonic devices. The electronic, optoelectronic and magneto-optic devices are based on monolayer / multilayer structures of thin films deposited through epitaxial processes onto single-crystalline substrates. The reliability, reproducibility, performance and lifetime of devices are determined by the purity of grown film, structural perfection, stoichiometry, surface flatness, interfaces and the homogeneity of the epitaxial layers. Certain applications require controlling the crystalline perfection and dopant concentration in the epitaxial layer. Epitaxy is discriminated in two different kinds:

(*i*) Homoepitaxy growth in which the epitaxial layer and the substrate are of the same material. If grown film and substrate have different lattice constants but same crystal lattices the film will be under strain with slightly different lattice constant than in its own bulk. Some novel properties may occur due to the electronic hybridization at the interface. Mostly commercial silicon epitaxy is Homoepitaxy.

(*ii*) Heteroepitaxy growth in which the epitaxial layer and substrate are of a different materials. The two crystal structure should be very similar if single-crystal growth is to be obtain and a large number of defects are to be avoided at the epitaxial-substrate interface such as $Al_xGa_{1-x}As$ on a GaAs substrate.

Epitaxial growth techniques have largely superseded the bulkgrowth for electronic circuit fabrication because the devices to be fabricated needs only upto few micron dimensions. The use of epitaxial growth therefore reduces the growth time, wafering cost and eliminates the wastages caused during growth, cleaning, cutting, polishing etc. The major advantage of the epitaxy is the uniformity in the composition, controlled growth parameters and better understanding of the growth itself.

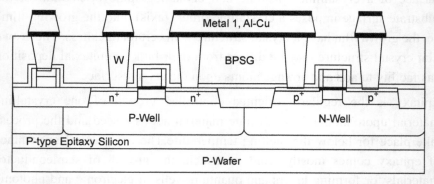

Fig. 2.1 Cross-sectional schematic of an epitaxial wafer used for CMOS integrated circuit fabrication.

Several epitaxial techniques are used to grow epitaxy layers of materials and compound semiconductors. The prominent among these techniques are

- Liquid Phase Epitaxy (LPE),
- Vapour Phase Epitaxy (VPE),
- Molecular Beam Epitaxy (MBE),
- Chemical Beam Epitaxy (CBE),
- Atomic Layer Epitaxy (ALE)

Few basic techniques have been listed in this text.

2.2 LIQUID PHASE EPITAXY

Liquid Phase Epitaxy (LPE) is a versatile, flexible method to grow thin layers of element as well as III-V, II-VI and IV-IV compounds for material investigations and device applications. Recent developments in the fabrication of the device structures like double heterostructure (DH) laser diodes, p-i-n photodiodes, avalanche photodiodes, Gunn diodes, integrated bipolar transistor-laser circuits, p-i-n FET photo-receivers, multi-quantum well lasers and rare earth doped injection lasers, have made liquid phase epitaxial technique, a unique distinction among the other epitaxial growth methods.

LPE means the growth of thin films from metallic solution on an oriented crystalline substrate. The solvent element can either be a constituent of the growing solid e.g. Ga or In, the solvent contains a small quantity of solute e.g. As in Ga or In, to epitaxial growth of epitaxial layers e.g. GaAs, InAs; which is transported towards the liquid-solid interface. LPE growth uses many principles, choice of solvents technological experience to growth of epitaxial layers. Epitaxial deposition can be done from concentrated solutions at higher temperature or from diluted solutions at low temperature or even it can be done from melts near the melting point.

LPE in practice prefer, dilute solutions at low temperature because it provide lower growth rates, improved thickness control, structural perfection and stoichiometry as well as reduce the detrimental effects of thermal expansion differences of substrate and epilayer. It also reduces the risk of unwanted spontaneously nucleated crystallites. LPE growth apparatus simply allows a growth solution of the desired composition to be placed with the substrate on contact for a certain time under controlled temperature conditions. In this technique, super-saturation necessary for deposition is achieved by reducing the temperature. For properly understanding the growth process, the relationship between the temperature and solubility as predicted by the phase diagram has to be utilized.

The process is best controlled if carrier transport occurs only by diffusion, i.e. the driving force in the solution is a concentration gradient of the solute. The growth boats shown in figure 2.2 are commonly designed such that essentially, only diffusion perpendicular to the interface occurs; convection and surface tension related transport are suppressed. The temperature gradient is minimized by utilizing larger dimension of the substrate compared to the height and radius of curvature. Applying these constraints, the LPE process can be treated as one-dimensional diffusion process where growth rate is found to be diffusion limited.

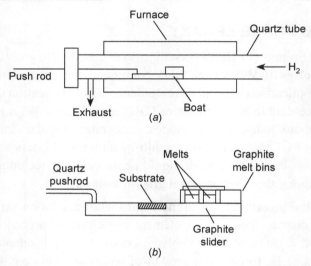

Fig. 2.2 (a) Growth furnace for LPE growth (b) Expanded schematic diagram of boat

There are three principal LPE growth techniques:

1. Tipping
2. Dipping
3. Sliding.

In the tipping technique, the substrate is held tightly at the upper end of a graphite boat and the growth solution is placed at the other end. By tipping the substrate, solution is brought into contact and then the epitaxial layer start growing on the substrate as the furnace cooled slowly. The solution remains in contact with the substrate between defined temperature interval and growth is dismissed by tipping the furnace reverse to its original position. The solution remaining on the film surface is removed by wiping and dissolving in a suitable solvent. The tipping process uses tipping furnace also known as horizontal furnace as shown in figure 2.3.

2.2 Liquid Phase Epitaxy

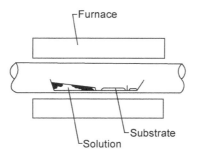

Fig. 2.3 Horizontal Furnace

The dipping technique uses a vertical furnace is shown in figure 2.4. The solution is contained in alumina or graphite crucible at the bottom end of the 3-zone furnace. The substrate fixed in a movable holder is initially positioned above the solution. At the desired temperature, growth is initiated by immersing the substrate in the solution and it is terminated by withdrawal of the substrate from the solution under the inert gas ambient. The apparatus used for the tipping and dipping techniques is very simple and easy to operate. However, growth of multiple layers by these techniques would require considerably more complex apparatus.

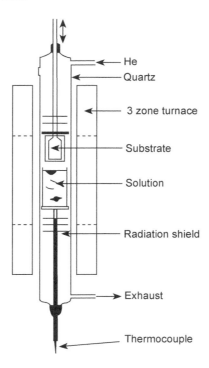

Fig. 2.4 Vertical Furnace

The third LPE technique known as the sliding technique uses a multibin graphite boat to grow multiple epitaxial layers. Figure 2.5 shows a LPE system with sliding technique. The principal components of this apparatus are a massive split graphite barrel with a graphite slider, a fused silica growth tube for providing a protective atmosphere and a horizontal resistance furnace. The graphite barrel has number of solution chambers depending on desired layers number to be grown, and the slider has two slots for the precursor seed substrate and the growth substrate. The substrate is brought into contact with the solutions by motion of the barrel over the slider so this operation can effortlessly be automated. The fused silica tube is usually within a heat pipe thermal liner in the furnace to ensure uniform temperature.

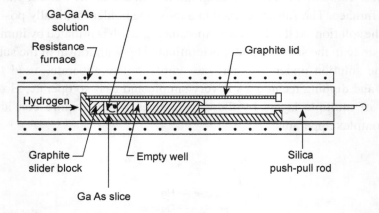

Fig. 2.5 LPE system with sliding technique

LPE has the advantages of low capital cost, high deposition rates, high material purity, no toxic gases as remnant and a relatively wide selection of dopants. Some disadvantages are listed as an inability to produce abrupt (monolayer) interfaces; poor large area uniformity; difficulty in varying stoichiometry and less controlling in the reproducibility of ternary III-V compounds. Advances in LPE equipment have allowed superlattice structures of 200-300 Å thick layers to be produced. Despite such progress LPE is not considered amenable to large scale high volume automated manufacturing.

2.3 VAPOR PHASE EPITAXY/CHEMICAL VAPOR DEPOSITION

Chemical reactions of gases utilize to form epitaxy layer on a hot surface. The change in free energy due to chemical reaction is reasonable force of deposition. Vapor phase epitaxy (VPE) of materials can be accomplished with different chemistries: hydride, halide or organometallic. The halide and hydride

2.3 Vapor Phase Epitaxy/Chemical Vapor Deposition

systems are common to the silicon semiconductor industry where epitaxial films are routinely deposited by the hydrogen reduction of chlorosilanes or the pyrolytic decomposition of silane. Chemical vapor deposition (CVD) is the most widely used form of VPE for Si epitaxy. CVD is a widely used materials-processing technology with majority applications involves applying solid thin-film coatings to surfaces, producing highly pure bulk materials and powders also generating composite materials via infiltration techniques. CVD is the construction of stable solids by decomposition of gaseous chemicals using heat, ultraviolet, RF, plasma, or a combination of sources.

CVD is quite old technology that was used to refine refractory metals in the 1800s as well as to produce filaments for Edison's incandescent carbon filament lamps in the early 1900s and for hard metal coatings in the 1950s. Semiconductor material preparation through CVD begins in the 1960s. At present the commercial silicon epitaxy production is accomplished largely by CVD using heat as the energy source to decompose gaseous chemicals. Radical changes in silicon material properties can be created over small distances within the same crystal so this capability permits the growth of lightly doped single crystal silicon on top of heavily doped single crystal silicon. CVD of single-crystal silicon is usually performed in a quartz reactor with susceptor is placed that provides mechanical support for the substrate wafers and uniform thermal ambience. Deposition occurs at a high temperature at which several chemical reactions take place when process gases flow into the chamber. So in CVD operation a precursor is introduced into a reaction chamber. Chamber is controlled by balanced flow regulators and control valves. Precursor molecules are drawn into the boundary layer passes through the substrate and get deposited on the surface of the substrate.

A common VPE process is shown in figure 2.6. The growth system is firstly cleaned by nitrogen or hydrogen for a short period of time then followed by a

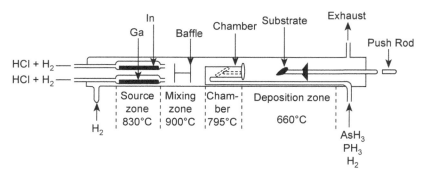

Fig. 2.6 A typical Vapor Phase Epitaxy (VPE) process

vapor HCl etching before the epitaxial layer deposition process. The deposition process is then initiated by directing the reactant gases into the reactor chamber where the preheated substrate is located.

The VPE growth method can be devided into the following steps and the flow diagram of the processing steps is shown in figure 2.7:

1. Introduction and transfer of the reactant species to the substrate region
2. Adsorption of the reactant species on the surface of substrate
3. Surface reaction such as surface diffusion, site accommodation, chemical reaction, and layer deposition take place on the substrate surface
4. Desorption of residual reactants and by-products from substrate surface
5. Transfer and removal of residual reactants and by-products from the substrate region

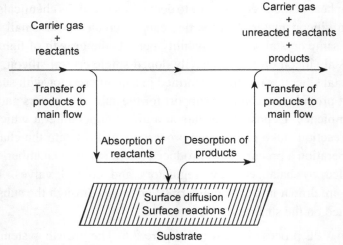

Fig. 2.7 Flow diagram of the sequence of steps in a VPE process

2.3.1 Growth Model and Theoretical Treatment

The type of fluid flow in a reactor is characterizes through Reynolds number (R_e):

$$R_e = \frac{D_r v \rho}{\mu} \qquad \ldots (2.1)$$

Where D_r is the diameter of the reaction tube, μ represents the gas viscosity, v denotes the gas velocity and ρ stands for gas density. Values of D_r and v are commonly several centimeters and tens of cm/s, respectively. The carrier gas is normally H_2, and by using typical values for ρ and μ, the value of R_e is about hundred.

2.3 Vapor Phase Epitaxy/Chemical Vapor Deposition

These parameters result in gas flow in continuous, regular, non-turbulent mode with a specified direction also known as laminar regime. Consequently, a boundary layer will form above the susceptor and at the walls of the reaction chamber due to reduced gas velocity at these interfaces. The thickness of the boundary layer, y, is defined as:

$$y = \left[\frac{D_r x}{R_e}\right]^{1/2} \quad \ldots (2.2)$$

Where x is the distance along the reactor.

Figure 2.8 shows the growth process of this boundary layer. Reactants are transported to the substrate surface and by-products of reaction diffuse back into the main gas stream across this boundary layer. The fluxes of species to and fro wafer surface are complex functions of the reactant, concentration, layer thickness and ambience such as temperature, pressure etc. The flux (J) is defined to be the product of D and dn/dy, by convention, and is the reactant flux of molecules per unit area per unit time, estimated as:

$$J = \frac{D(n_g - n_s)}{y} \quad \ldots (2.3)$$

Where D is the gas-phase diffusivity, which is function of pressure and temperature, n_g and n_s are the gas stream and surface reactant concentrations, respectively and y is the boundary layer thickness.

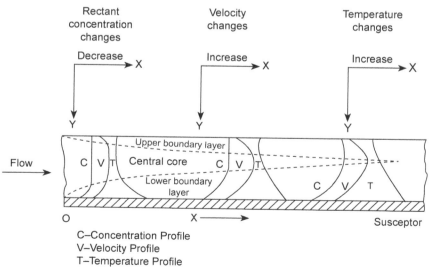

Fig. 2.8 Schematic of boundary layer formation

The reactant flux across the boundary layer is equal to the chemical reaction rate (k_s) in a steady state at the sampling surface. Therefore,

$$J = k_s n_s \qquad \ldots (2.4)$$

and

$$n_s = \frac{n_g}{1 + \dfrac{k_s y}{D}} \qquad \ldots (2.5)$$

The quantity D/y is often called the gas phase mass-transfer coefficient (h_g). So there are two limiting cases (1) when, n_s tends to zero so the reaction is limited by transport of reactant through the boundary layer. (2) when, $n_s \approx n_g$ so the surface chemical reaction rate dominates the growth process.

2.3.2 Growth Chemistry

The starting chemical for silicon epitaxy is silicon tetrachloride ($SiCl_4$). $SiCl_4$ has a lower reactivity than the other silicon hydrogen chloride compounds with respect to oxidizers in the carrier gas. Experimental results indicate the presence of many intermediate chemical species and residue. The overall reaction is:

$$SiCl_4 (g) + 2H_2 (g) \Leftrightarrow Si (s) + 4HCl (g)$$

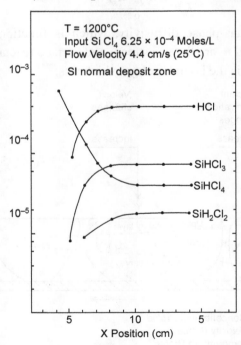

Fig. 2.9 Species detected by FTIR spectroscopy technique in a horizontal reactor using $SiCl_4$ and H_2.

A starting point in the analysis is to regulate the Si-Cl-H system, the equilibrium constant of each possible reaction and gaseous species partial pressures at the

2.3 Vapor Phase Epitaxy/Chemical Vapor Deposition

temperature of interest. Experimental calculations for equilibrium reveal fourteen species to be in equilibrium with solid silicon. In practice many of the species can be ignored because their partial pressures are less than 10^{-6} atm. The plot is for a particular Cl/H ratio (0.01), which is representative of the ratios that occur in epitaxial deposition. Utilizing FTIR spectroscopy technique in the reaction at 1200°C observes four species and plot of concentration versus position along the horizontal reactor is shown in figure 2.9. Detailed reaction mechanism is to be as:

$$SiCl_4 + H_2 \Leftrightarrow SiHCl_3 + HCl$$
$$SiHCl_3 + H_2 \Leftrightarrow SiH_2Cl_2 + HCl$$
$$SiH_2Cl \Leftrightarrow SiCl_2 + H_2$$
$$SiHCl \Leftrightarrow SiCl_2 + HCl$$
$$SiCl_2 + H \Leftrightarrow Si + 2HCl$$

The overall reaction rate can become negative i.e. etching occurs in lieu of deposition as above reactions are reversible.

Figure 2.10 shows an Arrhenius plot of growth rate, illustrating the overall reaction process. In region A the process can be characterized as reaction rate or kinetic limited i.e. one of the chemical reactions is the rate-limiting step, and is even reversible. Region B represents the situation in which the transport processes are rate limiting. The growth rate is limited by diffusing away the reaction products or by the amount of reactant reaching the wafer surface. This regime is termed mass transport or diffusion limited, and the growth gas. The slight increase of the growth rate at higher temperature in region B is due to the increased diffusivity of the species in gas phase with temperature. Industrial processes at atmospheric pressure are usually operated in region B to minimize the influence of temperature variation in the growth rate of uniform layer of epitaxy as seen in figure 2.10.

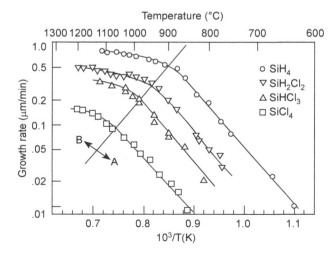

Fig 2.10 Arrhenius plot of growth rate

2.3.3 Doping

The impurity atoms hydrides are commonly used as the source of dopants during epitaxial growth. For instance,

$$2AsH_3 \text{ (g)} \longrightarrow 2As \text{ (s)} + 3H_2 \text{ (g)}$$
$$2As \text{ (s)} \longrightarrow 2As^+ \text{ (s)} + 2e^-$$

The incorporation process of dopant is illustrated schematically in figure 2.11.

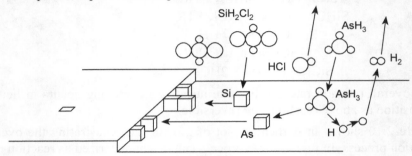

Fig. 2.11 Sketch of arsine doping incorporation and growth process

Some unintentional dopants are introduced in addition to deliberate dopants from the substrate through a process called autodoping. The dopant is released from the substrate through solid-state diffusion or evaporation and adjusted into the growing layer either by diffusion through the interface or through the gas. Autodoping is revealed as an enhanced region between the layer and the substrate. Figure 2.12 illustrate doping profile of an epitaxial layer detailing the various regions of autodoping, zone A occurs due to solid-state

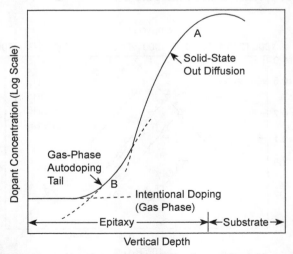

Fig. 2.12 Comprehensive doping profile of an epitaxial layer having the various regions of autodoping

2.3 Vapor Phase Epitaxy/Chemical Vapor Deposition

out-diffusion from the substrate and approximated by the complementary error function if the growth velocity is less than $2(D/t)^{1/2}$ (where D is the dopant diffusion constant and t denotes the deposition time. Zone B originates from gas-phase autodoping because evaporating of dopant from the wafer surface is supplied from the wafer interior by solid-state diffusion. The flux of dopant from an exposed surface decreases with time and once autodoping diminishes, the delibrated doping predominates and the profile becomes flat. Autodoping thus limits the minimum layer thickness that can be grown with controlled doping as well as the minimum dopant level.

Example 2.1

If the intrinsic diffusivity of boron in silicon is given by $D = 0.76 e^{\frac{3.46\, eV}{kT}}$ calculate the minimum growth rate that is required when a silicon epilayer is grown on a heavily boron-doped silicon substrate for 20 minutes at 1200°C in order that autodoping becomes insignificant. Discuss why a low growth temperature is essential to achieve a sub-micrometer thick silicon epitaxial layer.

Solution:

Diffusivity $\quad\quad D = 0.76 e^{\frac{3.46\, eV}{kT}}$

At 1200°C $\quad\quad D = 0.76 e^{\frac{3.46}{\frac{1.38 \times 10^{-23}}{1.6 \times 10^{-19}} \times 1473}}$

$= 0.76 e^{\frac{3.46}{0.127}}$

$= 0.76 \times 1.47^{-12} = 1.12 \times 10^{-12}$ (cm/s)

The minimum growth rate $> 2\left(\frac{D}{t}\right)^{1/2} = 2\left[\frac{1.12 \times 10^{-12}}{20 \times 10}\right]^{1/2} = 0.6$ (nm/s)

A low temperature is necessary to minimize the autodoping effect for submicron Si-epilayer deposition at reasonably low growth rates for tight thickness and doping control.

2.3.4 Reactors

Susceptors in epitaxial reactors are the analogs of crucibles in the crystal growing process. They provide mechanical support for the wafers and are the source of thermal energy for the reaction in induction-heated reactors. The geometric shape or configuration of the susceptor usually provides the name for reactor. There are three commercial epitaxial reactor designs: barrel, vertical, and horizontal, as shown in figure 2.13. In reactors, the reaction tube is either relatively cool during operation named as "Cold wall" or relatively hot named as "hot wall" reactor system. The growth rate is determined by mass transfer rate to the surface at high substrate temperatures and by chemical reaction rate on the surface at low substrate temperatures. The usual process for the CVD of polysilicon is hot wall operation. A water cooled coil is placed close to the susceptor so coupling can occur. The horizontal reactor is mostly used system because it offers high capacity and throughput but its drawback is nonuniform deposition over the entire susceptor. The disadvantage can be minimized by tilting the susceptor to 1.5-3° mitigate the non-uniformity substantially. In contrast, the vertical pancake reactor is capable of fine uniform growth with minimal autodoping problems but its disadvantages include mechanical complexity, low throughput, and susceptibility to particulate incorporation. Lastly the barrel reactor is an expanded version of the horizontal reactor in a different configuration. When used with a tilted susceptor, radiant-heated barrel reactors allow high-volume production and uniform grow.

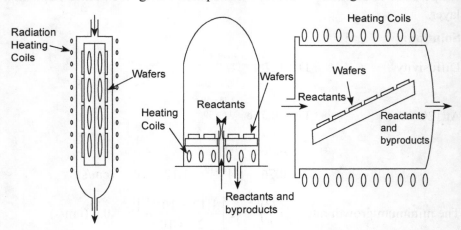

Fig. 2.13 Diagrams of three common reactors

2.4 DEFECTS

Epitaxial growth not only can introduce defects but also propagate defects. If these defects are in the active region of the wafer where the transistors are fabricated, they will often lead to device failures. These failures can be caused directly by electronic states associated with the defects, which lead to excessive leakage. Failures may also be less direct. The crystal perfection is frequently inferior of an epitaxial layer and can never exceeds that of the substrate. The crystal perfection is a function of the epitaxial process and the

2.4 Defects

properties of the substrate wafer itself. During processing, the defects may trap other impurities in the wafer that contribute to these electronic states. The defects may also lead to excessive impurity diffusion during processing, which changes the physical device structure.

Figure 2.14 illustrates some of the common structural defects in an epitaxial layer. Usually, defects can be reduced by a high operating growth temperature, high vacuum pressure, lower growth rate and cleaner substrate surface. A typical pre-epitaxy substrate cleaning process consists of a wet clean followed by a dilute HF dip and an in-situ HCl or SF_6 vapor etch.

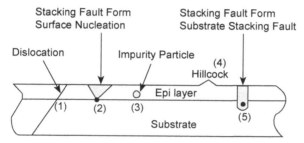

Fig. 2.14 Graphical representation of common defects occurring in epitaxial layers (1) line (or edge) dislocation firstly presented in the substrate then extended into the epitaxial layer, (2) epitaxial stacking fault nucleated by an impurity contamination on the substrate surface, (3) impurity precipitate caused by epitaxial process contamination, (4) growth spikes, and (5) bulk stacking faults, one of which intersects the substrate surface, thereby being extended into the layer

The most common types of defect in epitaxial silicon layers are dislocations and stacking faults. Dislocations (1) are extra or missing lines of atoms; stacking faults are an extra plane of atoms inserted into the crystal or a missing plane of atoms in the crystal. Dislocations are the 2-D analog of the stacking fault. Dislocations are fewer noticeable on the surface of the wafer but they are a serious yield concern. Dislocations may simply propagate from substrate dislocations. Stacking faults (2) in silicon normally occur in the <111> directions. With (100) wafers the stacking fault appears as a line along the <110> directions. Spikes (4) are protrusions from the epitaxial layer that show little or no alignment with the crystal directions it can be related to the onset of 3-D growth. Stacking faults and spikes often originate from a defect on the original wafer surface. These original defects include oxygen, metallic or alloy impurities, oxidation-induced on the wafer, and particles dumped on the surface of the wafer. Improvements in cleaning procedures of the wafers have dramatically reduced stacking fault densities in production silicon VPE.

Microscopic Growth Processes

The final points to consider in the CVD process are the conditions under which single-crystal films are obtained and the mechanism of their growth. A chemical reaction taking place and thereafter silicon atoms are start absorbing on the surface of the substrate. These atoms must migrate through the surface to find a crystallographically promising site where they can be incorporated into the lattice. At high growth rates, insufficient time is allowed for surface migration which results in polycrystalline growth of silicon epitaxy. The favorable sites are located at the leading edge of monolayer high steps. Thus, the growth is not vertical, but lateral. This effect accounts for the variation in growth rate with surface orientation, because the availability of sites and movement of steps is orientation dependent. The absorbed silicon atoms compete with dopant atoms, hydrogen, chlorine, and others atoms for these sites. Dopant atom concentration is usually low enough to be ignored, but impurities such as carbon affect the movement of silicon atoms on the surface and may nucleate a stacking fault or tripyramid defect. This lateral growth mechanism accounts for the effects that were discussed under pattern shift and distortion to off-orient <111> wafers.

2.5 TECHNICAL ISSUES FOR SI EPITAXY BY CVD

2.5.1 Uniformity/Quality

Uniformity and quality are the most important technological aspects of silicon epitaxy because device performance often depends directly upon these factors. Commercial systems avail thickness uniformity of ±2–4% and resistivity uniformity of ±4–10%. Such capability is acceptable for most device applications.

Statistical techniques are used to characterize variations in film properties. Average and standard deviation are well known concepts; however, results are often expressed as ±x–y% without specifying the definitions, as was done in the paragraph above. The uniformities noted above are for 90% of all points measured. Another specification frequently used is the total variation from maximum to minimum value, expressed as ±x–y% of the average. The maximum-minimum variation is easily calculated from the formula:

$$\pm \text{variation \%} = \frac{(\max - \min)}{(\max + \min)} \times 100$$

The term "90% of all data point" can be straight related to the standard deviation. Assuming there is a normal distribution of data, "90% of all data points" could be equal to 1.64 standard deviations. The maximum-minimum

2.5 Technical Issues for Si Epitaxy by CVD

variation is not directly related to statistical calculations; however, three standard deviations will normally include 99.6% of the data points.

Epitaxy quality includes crystal and surface defects, as well as metallic contamination.

2.5.2 Buried Layer Pattern Transfer

Buried layer pattern transfer is vital to the construction of bipolar integrated circuits. The buried layer (a pattern of heavily doped regions in the original substrate surface) is created in the wafer surface before epitaxial growth to provide low resistance ohmic paths for the collectors of the bipolar transistors. Often the reason for growing the epitaxial layer is to reduce a parasitic resistance. This is commonly done by growing a lightly doped epi layer on top of a heavily doped wafer or localized buried region. The transistor is formed in the epi layer, and the heavily doped region is essentially a buried contact to the bottom of the transistor. If the subsequent patterns are not aligned directly above these buried layer patterns, the transistors will not operate to specification.

Buried layer patterns in the wafer surface are created because the oxidation rate is higher in areas where the dopant concentration is higher. When the oxide is removed after buried layer diffusion, the higher oxidation rate leaves depressions in the wafer surface over each buried layer region. These depressions are used to align subsequent masks in the creation of an IC. An initial oxide is grown on a p-substrate and patterned for the guard ring diffusion. The guard rings are driven, the oxide is stripped, and a second oxide is grown. This oxide is patterned for the collector implant. The collector is implanted and driven, the oxide stripped, and a third oxide is grown. This oxide is patterned for the base implant. The base is implanted, the oxide stripped, the base is activated, and a fourth oxide is grown. This oxide is then patterned for the emitter implant. The emitter is implanted, the oxide is stripped, and a final oxide is grown. This oxide is patterned for the base contact (also known as the extrinsic base). A heavy p^+ implant is done, and the base contact and the emitter are activated in a final thermal step. A contact glass is deposited, contacts are made, and metallization layers are applied and patterned. As a final note, the extrinsic base implant mask can be eliminated by implanting the extrinsic base and intrinsic base at the same. If this implant is kept shallower and at a lower concentration than the emitter, the emitter diffusion will swamp the extrinsic base in the active region of the transistor device but will not affect it in extrinsic device region. The penalties that must be paid are a larger emitter–base capacitance, a lower breakdown voltage, and a higher base contact resistance. One of the first improvements in the basic 3-D technology is the addition of a buried collector. This is a heavily doped diffusion under the collector that shorts out the otherwise large collector series resistance. The use of buried collectors implies that the collector must be grown epitaxially on the substrate. This technology has been called Standard Buried Collector (SBC). Oxide-isolated SBC was the mainstay of the bipolar IC industry through the mid-1970s.

For Bipolar transistors it is necessary to align the upper device layers with this buried layer. To do this alignment marks may be etched into the substrate before epitaxial growth. Pattern shift is the tendency for position of the alignment marks to appear to move after the growth (Figure 2.15). The cause of pattern shift is the dependence of the growth rate on the exposed crystal orientation. Pattern shift is much less pronounced in (100) wafers than in (111). Generally, the lower chlorine content precursors (SiH_2Cl_2 and $SiHCl_3$) also show less pattern shift as does growth at higher temperature. The preferred method of measuring the film thickness is Fourier Transform Infrared (FTIR) spectroscopy, for relatively thick epitaxial layers, in which an infrared source is sent through a beam splitter to a movable mirror and to the surface of the wafer. The reflected radiation from both surfaces is added and sent to a detector. The distance of the mirrored path is cleared and monitoring of the intensity of the reflected beam as a function of the position of the mirror is done. The separation between these peaks is proportional to the thickness of the epitaxial layer.

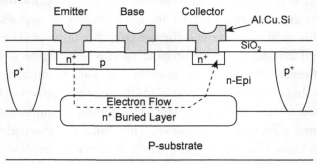

Fig. 2.15 Buried Layer Pattern

Because of different growth rates in different crystallographic directions, the buried layer patterns can be shifted relative to the region of high doping, and the pattern can be distorted or washed out. Pattern distortion is a change in size of the original pattern dimensions, often accompanied by sidewall fetching. Patterns can enlarge, shrink, facet and virtually disappear, depending upon the deposition conditions and orientation of the wafer. Pattern distortion occurs:

2.5 Technical Issues for Si Epitaxy by CVD

- With symmetrical pattern shift on off-orientation <111> surfaces, when the off-orientation is in the direction of the nearest (110) plane
- With unsymmetrical pattern shift on off-orientation (100) surfaces
- Without pattern shift on on-orientation (100) surfaces

Pattern distortion is summarized by:

Decreasing the chlorine concentration in the depositing gas

- Increasing the growth temperature
- Decreasing the growth pressure

Pattern shift is the motion of the surface pattern relative to the heavily doped buried layer region. The pattern shift ratio is the amount of shift divided by the thickness of the epitaxy layer. Pattern shift must be controlled to permit accurate registry of subsequent masks with the actual buried layer pattern. A zero pattern shift ratio is desirable but it is not achieved in commercial production except by reduced pressure deposition. A pattern shift ratios of 1:1 to 1.5:1 are commonly utilized in industries.

When a wafer is cut from a <111> oriented crystal, it is cut in a specific off-orientation direction, as depicted in figure 2.16. This orientation off the (111) plane toward the nearest (110) plane eliminates orange peel surface growth and provides for symmetric pattern shift during epitaxy growth. It is convenient to note that the symmetry of a dislocation etch pit reflects the symmetry of the off-orientation cut, as shown in figure 2.16. This observation provides a useful test for incorrect off-orientation wafers.

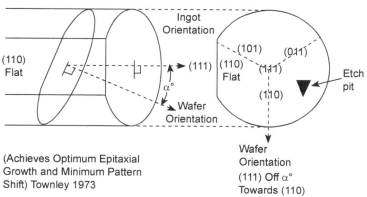

Fig. 2.16 Off-orientation <111> wafer cut from (111) crystal ingot.

For atmospheric pressure deposition, pattern shift is minimized by:

- Orienting (111) plane 3-5° toward the nearest (110) plane.
- Orienting the (100) plane parallel to the wafer surface within ±0.15°.

- Reducing the growth rate
- Increasing the growth temperature
- Reducing the growth pressure
- Reducing the chlorine content of the process gas

Fortunately, the conditions for reducing pattern shift are essentially the same as those for reducing pattern distortion, except that pattern distortion can be significant for on-orientation <100> surfaces.

SiH_4 offers the best results in terms of reduced pattern shift and distortion; however, the SiH_4 growth rate is normally less than 0.25 µm/min because of bell jar coating and gas phase reactions. For epitaxy layers thicker than about 1.5 µm, the best compromise is considered to be SiH_2Cl_2, operating at reduced pressure with growth rates in the 0.3-0.5 µm /min range. For epitaxy layers less than 1.5 µm, SiH_4 offers the advantage of lower temperature and the disadvantage of being more sensitive to oxidizer leaks. For lower growth pressures, pattern shift occurs to a lesser degree, becomes zero, and then becomes "negative," in that the pattern shift is in the direction opposite that for atmospheric pressure. The effect of pattern shift on pressure is illustrated in figure 2.17 where "negative" pattern shift was found for pressures below 100 torr at 0.3 µm /min using SiH_2Cl_2, at 1080°C in a cylinder reactor.

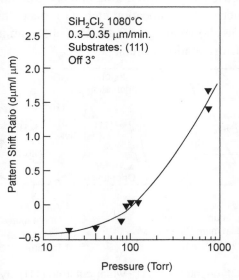

Fig. 2.17 Effect of pressure on pattern shift ratio.

Because of the "negative" pattern shift concept, process changes that "reduce" pattern shift must be measured in the algebraic sense. That is, if "negative" pattern shift is present, process changes that would "increase" pattern shift

2.5 Technical Issues for Si Epitaxy by CVD

would actually cause the "negative" shift to be closer to zero. Process changes that would "decrease" pattern shift will actually cause "negative" pattern shift to become more "negative" or further from zero. Pattern shift ratios versus pressure are provided in figure 2.18 for a radiantly heated cylinder and an induction heated vertical reactor. For pattern shift ratios above zero, decreasing the pressure and growth rate brings the shift closer to zero. For pattern shift ratios below zero, increasing the pressure and growth rate bring the shift closer to zero.

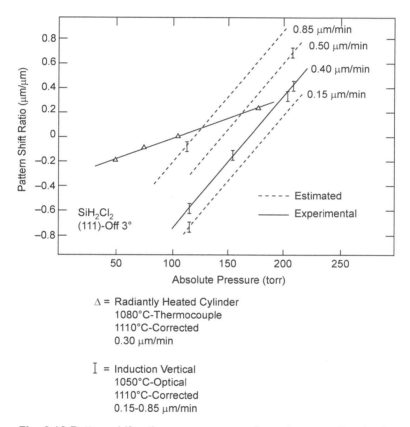

Fig. 2.18 Pattern shift ratio versus pressure for various growth rates in vertical and cylinder reactors.

Growth temperature has the strongest effect on pattern shift. A change of only 20-30°C has a major impact on the pattern shift ratio. Decreasing the deposition temperature causes the pattern shift ratio to increase in absolute terms; i.e., for "negative" pattern shift, decreasing the deposition temperature brings the pattern shift closer to zero.

2.6 AUTODOPING

Autodoping is the existence of unwanted dopant that contributed to the epitaxy layer by the wafers themselves. The intrinsic doping level is unwanted dopant contributed by the reactor parts. Autodoping are recognized of two types:
- Macro-autodoping in which dopant from the wafer surfaces both front and back surface contribute dopant generally to all the growing layers.
- Micro-autodoping in which dopant from one location migrates to another location on the same wafer.

Autodoping increases with increasing vapor pressure and increasing diffusion rates of the dopant. Sb causes the least autodoping, followed by As, B and P.

Macro-autodoping can be reduced by many techniques, including:
- The back of the wafer is sealed with oxide, nitride, or polycrystalline silicon fabricated by CVD process in the wafer preparation steps before epitaxy.
- Operating the reactor so that backside transfer of silicon occurs to seal the wafer's back during HCl etch. The process is effective but requires etch/coat time.
- Using a two-step process in which a undoped thin cap layer is deposited first, the system is then purged with frequency and the desired epitaxy layer is grown. This process is also effective for plug flow reactors where the gas composition can be quickly changed.
- Using a low/high temperature sequence in that the surface of the wafer is depleted of dopant through a high temperature bake then follow to a lower temperature epitaxy growth step. This process is mostly used for As.
- Decreasing the gas residence time by increasing the total flow rate and/or reducing the reactor pressure or volume.
- Operating at reduced pressure where the escaping tendency of the dopant on wafer surface and number of volume changes per minute is greatly increased (most effective for As).
- Reducing the concentration of dopant on the wafer surface by using ion implantation instead of dopant diffusion to dope the wafer surface prior to epitaxy growth (effective for all dopants). Micro-doping is depicted schematically in figure 2.19. Here, the n+ buried layer encroaches into the epitaxy layer by a combination of solid state diffusion and vapor transport during growth. Both vertical and lateral autodoping occur. Vertical autodoping is dopant moving vertically into the growing layer; lateral autodoping is movement to the sides of the buried layer region

2.7 Selective epitaxy

Micro-autodoping is reduced by the same techniques as macro-autodoping. Backside sealing is usually not required because IC processing is normally structured to avoid backside doping.

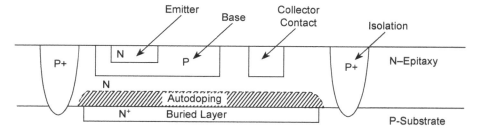

Fig. 2.19 Micro-autodoping in a buried layer bipolar device structure.

Pre-epitaxy oxidation can have a strong effect on autodoping. B is preferentially absorbed by a growing thermal oxide on silicon and the concentration near the single crystal surface is reduced. The opposite is true for P and As. If possible, pre-epitaxy doping for As and P should be accomplished with higher temperature, dry oxidation rather than lower temperature, wet oxidation. The opposite is true for B.

2.7 SELECTIVE EPITAXY

Selective epitaxy is the growth of single crystal silicon in windows etched into thermally grown oxide on the wafer surface. Selective epitaxy with or without controlled polycrystalline silicon being grown on top of the oxide is possible. Selective epitaxy without polycrystalline silicon overgrowth on the oxide is accomplished by depositing with conditions that do not promote nucleation. These conditions include depositing:

- With approximately a 3:1 ratio of chlorine to silicon, created by adding HCl to the deposit gas.
- At a slightly higher temperature to enhance the etching rate of the chlorine.
- At a reduced pressure where escape of volatile silicon chlorides is enhanced.
- A growth temperature of 1000°C using SiH_2Cl_2 with a 3/1 ratio of HCl/SiH_2Cl_2 at 100 torr is currently recommended for selective epitaxy without polycrystalline overgrowth.

Epitaxial lateral overgrowth is another variation on selective epitaxy in which the selectively grown epitaxy is permitted to grow up out of the windows and over the oxide surface. Single crystal layer on top of the oxide can be etched to disconnect it from the single substrate, thereby creating high quality single crystal islands on top of thermally grown oxide. Such structures have considerable promise for three-dimensional device structures.

Selective epitaxy with polycrystalline silicon overgrowth utilizes deposition conditions that promote nucleation, such as:

- The absence of chlorine.
- The use of SiH_4.
- The use of lower temperatures and higher pressures.

Selective epitaxy with polycrystalline silicon overgrowth can be accomplished with SiH_4 at 975°C and 760 torr deposition pressure. In device applications, the polycrystalline silicon overgrowth provides a place to create heavily doped contacts to the single crystal area without affecting the more lightly doped device regions.

2.8 LOW TEMPERATURE EPITAXY

Silicon epitaxy by CVD at lower temperatures is now of significant interest. Silicon epitaxy has been effectively demonstrated as:

- In the 800-1000°C range in a conventional reactor using SiH_4 or SiH_2Cl_2 after first initiating growth at a higher temperature.
- In the 800-1000°C range using SiH_4 and substituting He for H_2 as the main carrier gas.
- In the 850-1000°C range in a conventional reactor using SiH_4 with wafers that were etched in HF just prior to placing in the reactor.
- In the 850-900°C range in a conventional reactor using SiH_2Cl_2, at reduced pressures of 10-20 Torr.
- In the 800-900°C range using photo dissociation of silane and other silicon compounds.
- In the 750-950°C range in a cold wall reactor after removing the native oxide with a plasma etch.
- In the 700-900°C range by operating at very low pressure in a load locked, hot wall reactor.

All these low temperature epitaxy processes have limited application because they require very low growth rates to achieve even partially satisfactory crystal quality. As the required epitaxy layer thicknesses go below 0.6 µm, such techniques must be further developed and characterized for commercial production.

2.9 PHYSICAL VAPOR DEPOSITION (PVD)

PVD is a combined set of processes used to deposit thin layers of material, typically in the range of few nanometers to several micrometers. PVD is basically unlimited choice of coating materials: metals, alloys, semiconductors, metal oxides, carbides, nitrides, cermets, sulfides, selenides, tellurides etc. PVD process containing of three fundamental steps:

- Vaporization of the material from a solid source supported by high temperature vacuum or gaseous plasma.
- Transportation of the vapor in vacuum or partial vacuum towards the substrate surface
- Condensation onto the substrate to generate thin films

Physical Vapor Deposition (PVD) comprises of different methods, such as evaporation, sputtering, and molecular beam epitaxy (MBE)

- Evaporation: Material is heated to get gas phase below its melting temperature, where it then diffuses by high vacuum to the substrate.
- Sputtering: Plasma is created first which contains ions and electrons. Next, atoms from the target are ejected after being struck by ions. The atoms from the target then travel across the plasma and form a thin layer on the substrate.
- Molecular beam epitaxy (MBE): The substrate is cleaned then loaded into a chamber that is evacuated and heated to drive off surface contaminants and to roughen the surface of the substrate. The molecular beams emit a small amount of source material through a shutter that is collects on the substrate.

PVD is used in a variety of applications, including fabrication of interconnects, microelectronic devices, diffusion barriers, optical and conductive coatings, surface modification, battery and fuel cell electrodes. We will discuss detailed PVD application in Metallization chapter (Chapter 8). Here we will focus on MBE process of epitaxy.

2.9.1 Molecular Beam Epitaxy (MBE)

MBE, a technique of vacuum evaporation is one of the easiest and oldest techniques of depositing solid films. Although vacuum evaporation was used as early as in the 1950s for preparing semiconductors, epitaxial growth conditions were not realized until improvements occurred in Ultra-High Vacuum (UHV) technology, in the design and control of the sources and substrate cleaning procedures. MBE has now become a versatile technique for growing epitaxial thin films of semiconductors, metals and superconductors.

A functional schematic diagram of a MBE system is shown in figure 2.20. It consists of a growth chamber and auxiliary chamber (not present with first generation systems), diffusion pumps and a load-lock. Each chamber has an associated pumping system. The load-lock facilitates the introduction and removal of samples or wafers without significantly influencing the growth chamber vacuum. The auxiliary chamber may contain supplementary surface analytical tools not contained in the growth chamber, additional deposition equipment and other processing equipment. Separating equipment in this manner allows for more efficacious use of the growth chamber and enhances the quality of operations in both the auxiliary and growth chambers.

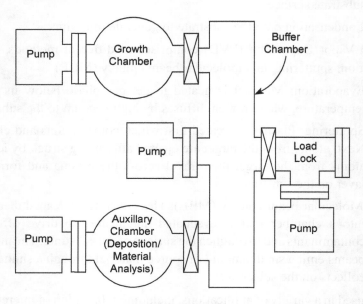

Fig. 2.20 MBE System

The growth chamber is shown in greater detail in figure. 2.21. Its main elements are: sources of molecular beams; a manipulator for heating, translating and rotating the sample; a cryoshroud surrounding the growth region; shutters to occlude the molecular beams; a nude Bayard Alpert gauge to measure chamber base pressure and molecular beam fluxes; a RHEED (Reflection Electron Diffraction) gun and screen to monitor film surface structure and a quadrupole mass analyzer to monitor specific background gas species or molecular beam flux compositions. The auxiliary chamber may be host to a wide variety of process and analytical equipment. Typical surface analytical equipment would be: an Auger electron spectrometer, or equipment for Secondary Ion Mass Spectrometry (SIMS), ESCA (Electron Spectroscopy for Chemical Analysis) or XPS (X-ray Photoelectron Spectroscopy). There may be a heated sample

2.9 Physical Vapor Deposition (PVD)

station and an ion bombardment gun for surface cleaning associated with this equipment. Process equipment may include sources for deposition or ion beam etching.

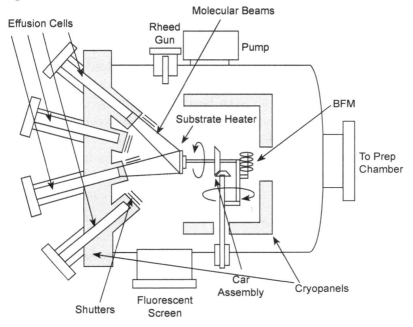

Fig. 2.21 Schematic cross-section of a typical MBE growth chamber.

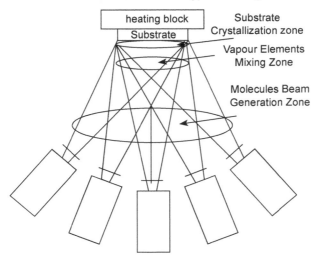

Fig. 2.22 Effusion Cell

The principle underlying MBE growth is quite simple: it basically consists of atoms or clusters of atoms, which are produced through heating up a solid source. They then move in an UHV environment and impose on a hot substrate surface, where they can diffuse and eventually incorporate into the growing film. The MBE growth process involves controlling to achieve epitaxial growth via shutters, source temperature, molecular and/or atomic beams directed at a single crystal sample (suitably in-situ heated). The beams are thermally generated in Knudsen-type effusion cells (shown in figure 2.22) that contain the essential elements or compounds of the desired epitaxial films. The temperatures of the cells are precisely controlled to give the thermal beams of appropriate intensity. The beam fluxes emerging from these non-equilibrium effusion cells are generally determined experimentally in most cases using movable nude ionization gauge placed in the substrate position. The cells are made from Pyrolytic Boron Nitride (PBN) or high purity graphite materials, those are non-reactive, refractory materials that can withstand high temperatures and strictly do not contribute to the molecular beams.

The cell consists of an inner crucible and an outside tube which is wound with Ta or Mo wires for resistive heating. The various cells are all placed and angled in such a way that their beams converge on the substrate for epitaxial growth. A chemically stable W-Re thermocouple facilitates precise control of the cell temperature which is very essential for achieving constant growth rates since small temperature fluctuations of the order of $\pm 1°C$ can result in ± 2 to 4 percent fluctuations in molecular beam intensity. Individual shutters provided for each cell and the cell temperature can be computer controlled to achieve high reproducibility with little human interventions. The cells are individually surrounded by a liquid nitrogen covering to prevent cross heating and cross contamination. For group V elements, a high temperature cracker which dissociates the tetramers to dimers, with internal buffer is incorporated at the exit end of effusion cell. The gas background necessary to minimize unintentional contamination is predicated by the relatively slow film growth rate of approximately $1 \mu m/h$ and is usually in the 10^{-11} torr range. At this pressure, the mean free path of gases in the beams themselves is several orders of magnitude greater than the normal source-to-sample distance that of about 15 cm. Hence, the beams impinge unreacted on the sample with a cryo-shroud cooled by liquid nitrogen. Reactions take place predominantly at the substrate surface where the source beams are incorporated into the developing film. Proper initial preparation of the substrate will present a clean, single crystal surface upon which the developing film can deposit epitaxially. Actuation properly and timely of the source shutters allows film growth to be controlled to the monolayer level. Monolayers level formation ability and precisely control epitaxial film growth and composition has attracted the attention of material and device scientists towards MBE.

2.9 Physical Vapor Deposition (PVD)

Silicon MBE is performed under UHV conditions of 10^{-8} to 10^{-10} torr. The mean free path of the atom is given by $5 \times 10^{-3}/P$ where P is the system pressure in Torr. Transport velocity is dominated by thermal energy effects at a typical pressure of 10^{-9} torr and L is 5×10^6 cm. The lack of intermediate reactions and diffusion effects, coupled with relatively high thermal velocities, results in film properties changing rapidly with any change of the source. The typical growth temperature in order to reduce autodoping and out-diffusion is between 400°C and 800°C. Growth rates are in the range of 0.01 to 0.3 µm/minute.

Fig. 2.22 MBE Chamber

Molecular Beam Epitaxial growth technique has a number of advantages over other techniques. A particular advantage is that it permits growth of crystalline layers at temperatures where solid-state diffusion is negligible. Since chemical decomposition is not required for growth, deposition species need require only enough energy to migrate along the substrate surface to crystalline bonding site. The impurity dopant incorporation during molecular beam epitaxial growth is possible by having an additional source of the dopant. As a result, MBE has rapidly established itself as a versatile technique for growing elemental and compound semiconductor films. Thus using MBE, it is possible to produce multilayered structures including superlattices with layer thickness as low as 10 Å for DH lasers and waveguide applications.

However, there are few limitations in the epitaxial growth of compound semiconductors by MBE technique. The ultra-high vacuum apparatus is very expensive. Frequent shutdowns are required to replenish the source materials and opening the UHV apparatus. A major problem is the difficulty in growing phosphorus-containing materials.

2.10 SILCON-ON-INSULATOR (SOI)

Silicon device structures have integral problems that are related to parasitic circuit elements arising from junction capacitance. These effects become more severe as device dimensions shrink. In SOI technology, different parts of the device circuit are built on separate small islands of silicon that are fabricated over insulating substrates for providing a certain degree of isolation between circuits on different islands. Conventional wafer fabrication techniques are used for necessary interconnections between these isolated circuits. The introduction of SOI technology in IC fabrication improves speed in complex dense circuits and especially device's electrical performance by reducing parasitic capacitances. It also increases radiation hardness for aerospace applications. Figure 2.23 shows bulk and SOI CMOS devices.

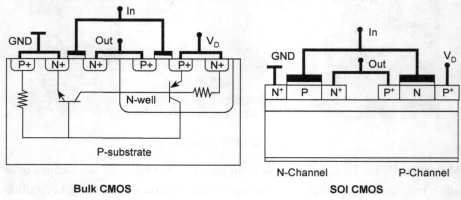

Fig. 2.23 Bulk and SOI CMOS

A practical means to avoid the problem of parasitic junction capacitances is to fabricate devices in small islands of silicon on an insulating substrate. SOI technology is not new but its studies and practicing starts as early as the 1960's. There are numerous techniques through which SOI implementation may be achieved but earlier work dominantly involves Silicon-on-Sapphire (SOS). High quality strailess SOS epitaxial layers can be fabricated as the lattice parameters of silicon and sapphire are quite similar. Still, the high cost of sapphire substrates, low yield, and lack of commercially sustainable applications limit the use of SOS to primarily military applications. Sidewise from SOS, techniques for growing single-crystal silicon on several insulating substrates have also been discovered and successfully realized but most of these processes never became broadly used in IC fabrication. The other alternative Silicon-on-Insulator (SOI) approach is SIMOX (Separation by Implantation of Oxygen) that utilizes high dose blanket oxygen ion implantation to form a sandwiched buried oxide layer to isolate devices from the wafer substrate. Another exciting approach is wafer bonding which utilizes Van der Waals forces to bond two polished silicon wafers, at least one of them is covered with thermal oxide at about 1000°C in a very clean environment. More recent approaches include the combination of layer cleavage and wafer bonding using hydrogen or helium ion implantation (ion-cut) as well as epitaxial growth on porous silicon and wafer bonding.

2.11 SILICON ON SAPPHIRE (SOS)

SOS involves the epitaxial growing thin layer of silicon on insulating substrate of sapphire (Al_2O_3) at high temperature as shown in figure 2.24. This growth is labeled 'hetero-epitaxy', as the grown material layer is different from that of the base material or substrate. Nonetheless, the materials and equipment utilized for the hetero-epitaxial growth of SOS are basically identical to those used in homo-epitaxial growth. Its main advantage for electronic circuits is the highly insulating sapphire substrate that provide very low parasitic capacitance, hence increased speed, lower power consumption, better linearity and more isolation than bulk silicon.

Silane (SiH_4) is mostly used as source of silicon for SOS growth. Its pyrolysis reaction in a carrier hydrogen gas is given as

$$SiH_4\,(g) \longrightarrow Si\,(s) + 2H_2\,(g)$$

Results in the fabrication of silicon layer over the sapphire base. The deposition temperature is retained below 1050°C to prevent the auto-deposition of aluminum element from the sapphire substrate into the silicon layer. The desired <100> silicon orientation has been realized on various sapphire orientations, i.e., <1102>, <0112>, <1012>. The r-plane of sapphire has oxygen atoms spaced at a close distance to the spacing of the atoms in the (100) plane of a silicon crystal as well as it has square symmetry that also mirrors the symmetry of the (100) plane of silicon. SOS has some inherent drawbacks as in every technology which needs to be addressed before its benefits can be realized. Lattice parameter mismatch between the grown silicon layer and the sapphire substrate frequently generates misfit dislocations, edge dislocations and stacking faults in SOS devices, with the defect density varying inversely with the distance from the substrate. The coefficients of thermal expansion difference between the silicon and sapphire also results in a residual stress within the silicon layer also tends to reduce hole mobility. This, coupled with the lower hole and electron mobilities caused by defects, ultimately results in poorer performance of MOS devices by SOS in contrast to those fabricated on bulk silicon.

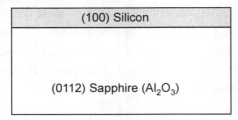

Fig. 2.24 Cross Sectional View of Silicon on Sapphire

2.12 SILICON ON SIO$_2$

Current SOI technology is silicon on SiO_2 used in Bipolar, MOS and FinFET devices with excellent properties that involves the use of buried oxide layers, which are basically subsurface layer of SiO_2 formed through ion implantation of very high doses of oxygen.

SOI fabrication using buried oxide layers follows basic steps such as:

- O_2 is implanted into the silicon substrate at a high dosage (approx. 2×10^{18} cm^{-2}) and energy (150-300 keV).
- Post deposition annealing is used to achieve restoration of the crystallinity of the substrate surface and formation of the buried oxide itself at high temperature (1100-1175°C) in an inert environment (e.g., using N_2) for 3-5 hours.
- A layer of epitaxial silicon, that serve as the layer over which the circuits will be built, is deposited over the buried oxide.

Recently, buried silicon nitride (Si_3N_4) layers have equally been successfully used in SOI technology.

2.13 SUMMARY

Epitaxy process is a integral part of circuit manufacturing. This chapter starts discussion with homo-epitaxial and herto-epitaxial processes. Homo-epitaxial silicon structures remain the popular design choice but SIO and SOS are undoubtedly receive considerable research attention. The epitaxial growth of silicon from the gas vapor is a well-established process for fabricating advanced semiconductor technologies. An understanding of the process requires a detailed knowledge of both the gas phase chemistry and the events at the surface.

The second half of the chapter presented growth techniques that have been developed for controllably producing thin epitaxial layers. Several epitaxial techniques have been used for the growth of epilayers, the prominent among these techniques are Liquid Phase Epitaxy (LPE), Vapor Phase Epitaxy (VPE), and Molecular Beam Epitaxy (MBE), have discussed. The common techniques for thin epitaxial growth are Chemical Vapor Deposition (CVD) and Molecular Beam Epitaxy (MBE) where CVD is chemical deposition process and MBE is physical deposition process. In CVD gases and dopants are transported in vapor form to the substrate, where a chemical reaction occurs that results in the deposition of the epitaxial layer. MBE is done by the evaporation of a species in an ultrahigh-vacuum system. It is a low-temperature process so it has low growth rate. MBE can grow single-crystal, multilayer structures with dimensions on the order of atomic layers.

PROBLEMS

1. Determine the amount of mask compensation needed for an epitaxial wafer of <150> orientation containing an antimony buried layer with an epitaxial thickness of 6 μm.

2. In a 1 h process at 1000°C using dichlorosilane, a 10μm layer is grown on 25 substrate of 150 mm diameter in a horizontal reactor. Adopt growth rate of 1.5 μm/minutes.

3. Calculate the number of liters of hydrogen at STP that would have to be supplied into the reactor for process of problem 2. What do you determine?

4. Compare liquid phase epitaxy and vapour phase epitaxy. Which one is more suitable and why?

5. What are doping and anti doping? Explain their role in semiconductor manufacturing.

REFERENCES

1. H. Manasevit and R. Simpson, "A Survey of the Heteroepitaxial Growth of Semiconductor Films on Insulating Substrates," *J. Cryst. Growth* 22:125 (1974).
2. P.K. Vasudev, "Silicon-on-Sapphire Heteroepitaxy," Epitaxial Silicon Technology, Academic Press, Orlando, FL, (1986).
3. W.I. Wang, "Molecular Beam Epitaxial Growth and Materials Properties of GaAs and AlGaAs on Si (100)," *Appl. Phys. Lett.* 44:1149 (1984).
4. W.T. Masselink, T. Henderson, J. Klem, R. Fischer, P. Pearah, H. Morkoc, M. Hafich, P. D. Wang, and G. Y. Robinson, "Optical Properties of GaAs on (100) Si Using Molecular Beam Epitaxy," *Appl. Phys. Lett.* 45:1309 (1984).
5. T. Soga and S. Hattori, "Epitaxial Growth and Material Properties of GaAs on SiGrown by MOCVD," *J. Cryst. Growth* 77:498 (1986).
6. T. Soga, S. Hattori, S. Sakai, M. Takeyasu, and M. Umeno, "MOCVD Growth of GaAs on Si Substrates with AlGaP and Strained Layer Superlattice Layers," *J. Appl. Phys.* 57:4578 (1985).
7. R. People and J.C. Bean, "Calculation of Critical Layer Thickness Versus Lattice Mismatch for GexSi1-xAs: Strained-Layer Heterointerfaces," *Appl. Phys. Lett.* 47:322 (1985); 49:229 (1986).
8. A.J. Shuskus, T.M. Reeder, and E.L. Paradis, "rf-sputtered aluminum nitride films on sapphire," *Appl. Phys. Lett.* 24:155 (1974).
9. G.B. Stringfellow, "Organometallic Vapor-Phase Epitaxy," Academic Press, Boston, (1989).
10. R.M. Lum, J.K. Klingert, and M.G. Lamont, "Comparison of Alternate As-sources to Arsine in the MOCVD Growth of GaAs," in Fourth Int. Conf MOVPE, 1-3 (1988).
11. C.H. Chen, C.A. Larsen, G.B. Stringfellow, D.W. Brown, and A.J. Robertson, "MOVPE Growth of InP Using Isobutylphosphine and tert-Butylphosphine," *J. Cryst. Growth* 77:11 (1986).
12. R.J. Field and S.K. Ghandhi, "Doping of GaAs in a Low Pressure Organometallic CVD System," *J. Cryst. Growth* 74:543 (1986).
13. T.F. Kuech, "Metal-Organic Vapor Phase Epitaxy of Compound Semiconductors," *Mater. Sci. Rep.* 2:1 (1987).
14. D.J. Schlyer and M.A. Ring, "An Examination of the Product-Catalyzed Reaction of Trimethylgallium with Arsine," *J. Organometall. Chem.* 114:9 (1976).

References

15. D.H. Reep and S.K. Ghandhi, "Deposition of GaAs Epitaxial Layers by Organometallic CVD," *J. Electrochem. Soc.* 130:675 (1983).
16. P. Rai-Chaudhury, "Epitaxial Gallium Arsenide from Trimethylgallium and Arsine," *J. Electrochem. Soc.* 116:1745 (1969).
17. Y. Seki, K. Tanno, K. Iida, and E. Ichiki, "Properties of Epitaxial GaAs Layers from a Triethylgallium and Arsine System," *J. Electrochem. Soc.* 122:1108 (1975).
18. T.F. Kuech, M.A. Tischler, P.J. Wang, G. Scilla, R. Potemski, and F. Cardone, "Controlled Carbon Doping of GaAs by Metalorganic Vapor Phase Epitaxy," *Appl. Phys. Lett.* 53:1317 (1988).
19. B.T. Cunningham, M.A. Haase, M.J. McCollum, J.E. Baker, and G.E. Stillman, "Heavy Carbon Doping of Metalorganic Chemical Vapor Deposition Grown GaAs Using Carbon Tetrachloride," *Appl. Phys. Lett.* 54:1905 (1989).
20. B.T. Cunningham, L.J. Guido, J.E. Baker, J.S. Major, N. Holonyak and G. E. Stillman, "Carbon Diffusion in Undoped n-type and p-type GaAs," *Appl. Phys. Lett.* 55:687 (1989).
21. M.A. Tischler, "Advances in Metalorganic Vapor-Phase Epitaxy," *IBM J. Res. Dev,* 34:828 (1990).
22. T.F. Kuech, E. Veuhoff, D.J. Wolford, and J.A. Bradley, "Low Temperature Growth of $Al_xGa_{1-x}As$ by MOCVD," GaAs, Related Compounds, *11th Int. Symp.*, p. 181 (1985).
23. J.R. Shealey and J.M. Woodall, "A New Technique for Gettering Oxygen and Moisture from Gases in Semiconductor Processing," *Appl. Phys. Lett.* 68:157 (1984).
24. J.F. Gibbons, C.M. Gronet, and K.E. Williams, "Limited Reaction Processing: Silicon Epitaxy," *Appl. Phys. Lett.*, 47:721 (1985).
25. S.A. Campbell, J.D. Leighton, G.H. Case, and K. Knutson, "Very Thin Silicon Epitaxial Layers Grown Using Rapid Thermal Vapor Phase Epitaxy," *J. Vacuum Sci. Technol. B*, 7:1080 (1989).
26. M.L. Green, D. Brasen, H. Luftman, and V.C. Kannan, "High Quality Homoepitaxial Silicon Films Deposited by Rapid Thermal Chemical Vapor Deposition," *J. Appl. Phys.*, 65:2558 (1989).
27. T.Y. Hsieh, K.H. Jung, and D.L. Kwong, "Silicon Homoepitaxy by Rapid Thermal Processing Chemical Vapor Deposition," *J. Electrochem. Soc.*, 138:1188 (1991).

28. K.L. Knutson, S.A. Campbell, and F. Dunn, "Three Dimensional Temperature Uniformity Modeling of a Rapid Thermal Processing Chamber," *IEEE Trans. Semicond. Manuf* 7(1):68 (1994).

29. B.S. Mayerson, "Low-Temperature Silicon Epitaxy by Ultrahigh Vacuum/Chemical Vapor Deposition," *Appl. Phys. Lett.*, 48:797 (1986).

30. B.S. Mayerson, E. Ganin, D.A. Smith, and T.N. Nguyen, "Low Temperature Silicon Epitaxy by Hot Wall Ultrahigh Vacuum/Chemical Vapor Deposition Techniques: Surface Optimization," *J. Electrochem. Soc.*, 133:1232 (1986).

31. B.S. Meyerson, "Low Temperature Si and Ge:Si Epitaxy by Ultrahigh Vacuum/Chemical Vapor Deposition: Process Fundamentals," *IBM J. Res. Dev.*, 34:806 (1990).

32. J. Davies and D. Williams, "III-V MBE Growth System, in The Technology and Physics of Molecular Beam Epitaxy," Plenum, New York (1985).

33. D. Bellevance, "Industrial Application: Perspective and Requirements, in Silicon Molecular Beam Epitaxy 2," CRC, Boca Raton, FL, 153 (1985).

34. T. Tatsumi, H. Hirayama, and N. Aizaki, "Si Particle Density Reduction in Si Molecular Beam Epitaxy Using a Deflection Electrode," *Appl. Phys. Lett.*, 54:629 (1989).

35. A. Von Gorkum, "Performance and Processing Line Integration of a Silicon Molecular Beam Epitaxy System," *Proc. 3rd Int. Symp. Si MBE, Thin Solid Films*, 184:207 (1990).

36. K. Fujiwara, K. Kanamoto, Y.N. Ohta, Y. Tokuda, and T. Nakayama, "Classification and Origins of GaAs Oval Defects Grown by Molecular Beam Epitaxy," *J. Cryst. Growth*, 80:104 (1987).

37. A.Y. Cho, "Advances in Molecular Beam Epitaxy (MBE)," *J. Cryst. Growth* 111:1 (1991).

38. J. Saito, K. Nanbu, T. Ishikawa, and K. Kondo, "In situ Cleaning of GaAs Substrates with HCl Gas and Hydrogen Mixture Prior to MBE Growth," *J. Cryst. Growth* 95:322 (1989).

39. N. Chand, "A Simple Method for Elimination of Gallium-Source Related Oval Defects in Molecular Beam Epitaxy of GaAs" *Appl. Phys. Lett.*, 56:466 (1990).

40. J.H. Neave, P. Blood, and B.A. Joyce, "A Correlation between Electron Traps and Growth Processes in n-GaAs Prepared by Molecular Beam Epitaxy," *Appl. Phys. Lett.,* 36:311 (1980).

References

41. W.K. Burton, N. Cabrera, and F.C. Franks, "The growth of crystals and the equilibrium structure of their surfaces," *Philos. Trans. R. Soc. London, Ser.* A 243:299 (1951).
42. M.G. Legally, "Atoms in Motion on Surfaces," *Phys. Today* 46:24 (1993).

❏❏❏

3

Oxidation

3.1 INTRODUCTION

Silicon is unique among semiconductor materials and what makes it so popular is that silicon forms an excellent native oxide, SiO_2 with ease. This oxide is widely used as an insulator both in active devices such as MOSFETs and in the region between the active devices, known as the field.

It is privileged that silicon forms protective oxide easily, otherwise we should have to depend upon deposited insulators.

In oxidation process, a semiconductor or metal is converted to an oxide. In technology, oxidation of various materials plays a role, but conversion of parts or fully of a semiconductor wafer into SiO_2 is chief oxidation process. At high temperature, to generate SiO_x, the chemical reaction between oxygen and silicon takes place. Besides this high temperature requirement, a shallow layer of native oxide, approximately 0.5 nm to 2 nm thick can be grown. But this shallow layer is not suitable for most applications. The heavy layer is grown by consuming the underlying Si to form SiO_x. This is known as grown layer. A chemical vapor deposition (CVD) process using Si and O precursor molecules is also use to grow SiO_x. This layer is known aa deposited layer.

One major function of SiO_x is to protect the wafer from contamination, both physical and chemical. The oxide layer does not allow dust from interacting with wafer and protects the wafer surface from scratches. Thus, it helps in contamination minimization. Other than this, oxide layer also protects the wafer from chemical impurities. For doping SiO_x acts as a hard mask and during patterning as an etch stop. SiO_x was also used for separating different metallization layers, though this is usually a deposited layer.

To avoid induced charge due to the metal layers, also oxide layer is deposited. Then, this layer is known as field oxide. For different applications, different thickness of the oxide layer is required. These are shown in figure 3.1.

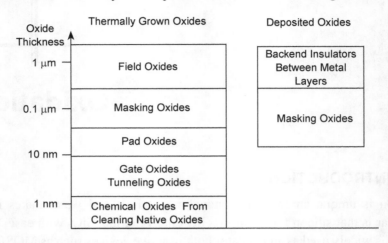

Fig. 3.1 Uses of SiO_2 in silicon based IC technology.

Some of the properties of SiO_2 are listed in the table 3.1:

Table 3.1 Physical & chemical properties of Silicon

Properties	SiO_2
Crystal Structure	Amorphous
Melting Point (°C)	1700 appx.
Density (gm/cm³)	2.2
Dielectric Constant	3.9
Refractive Index	1.46
Dielectric Strength (V/cm)	10^7
Energy gap at 300 K(eV)	9
DC resistivity at 25°C (ohm-cm)	10^{14}–10^{16}
Thermal Conductivity at 300K (W/cm-degree K)	0.014
Infrared absorption bend (μm)	9.3
Atomic Weight	60.08 g/mole
Molecules	$2.3*10^{22}$/cm³
Specific Heat	1.0 J/g-K
Thermal Expansion Co-efficient	$5.6*10^{-7}$/K
Young's Modulus	$6.6*10^6$ N/m²
Poisson's Ratio	0.17

Application

Following are some important applications of silicon dioxide:
1. Quartz is used in the glass industry as a raw material for manufacturing glass.
2. Silica is used as a raw material for manufacturing concrete.
3. Silica is added to varnishes because of its hardness and resistance to scratch.
4. Amorphous silica is added as fillers to the rubber during the manufacturing of tires. This helps reduce the fuel consumption of the vehicle.
5. Silica is used to produe silicon.
6. As silica is a good insulator, it is used as a filler material in electronic circuits.
7. Quartz has piezoelectric properties so it is also used in transducers.
8. Its ability to absorb moisture is utilised by using as a desiccant.

Roles

The SiO_2 layer has multipurpose role on chip. Si is the only semiconductor material which has a native oxide. This native oxide achieve all the properties of SiO_2. The SiO_2 role in semiconductor fabrication can be described as:
1. During diffusion on ion implantation process, it provides mask role for dopants. It acts as a diffusion mask and allows selective implantation into silicon wafer. This is achieved via the window etched into oxide.
2. Oxide layer provides surface passivation. It creates a shield of SiO_2 layer on the wafer surface. It defends the junction from moisture.
3. Another feature is device isolation. It forms an insulation layer on the water surface.
4. As component in MOS structures (gate oxides). In MOS devices silicon dio-oxide acts as the active gate electrode.
6. In multi-level metallization systems, it provides electrical isolation.
7. High thermal conductivity
8. No diffusion of Cu or other ions into dielectric
9. No leakage between conductors

3.2 GROWTH AND KINETICS

The temperature is key parameter to control oxide growth during SiO_2. The oxidation environment determines the growth rate, depending on whether it is wet (H_2O) or dry (O_2). It also play an important role in the crystal orientation of the wafer.

To deposit thicker oxides, several methods are used:
- Thermal Oxidation
- Electrochemical Oxidation.

Silicon is consumed as oxide grows, and the resulting oxide expands during growth so movement of the interface is as shown in figure 3.2.

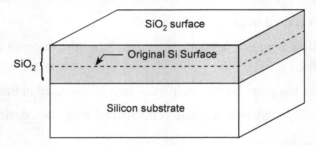

Fig. 3.2 Lateral view of SiO_2 growth on silicon substrate

As the underlying Si is consumed, the SiO_2 interface travels deeper into the wafer. Thus, we obtain the higher thickness as compared to the initial Si thickness. A oxide layer silicon interface is shown in figure 3.3.

Here,   d = thickness of the original Si layer

Si has a density of 2.33 gcm^{-3} (ρ_{Si}) and an atomic weight of 28.08 gmol^{-1} (Z_{Si}) while SiO_2 has a density of 2.65 gcm^{-3} (ρ_{SiO_2}) and a molecular weight of 60.08 gmol^{-1} (Z_{SiO_2}).

Given that the cross section area, A, is the same it is possible to use the law of molar conservancy to derive the relation between d and D. This can be represented as

$$\frac{dA_{\rho Si}}{Z_{Si}} = \frac{DA_{\rho SiO_2}}{Z_{SiO_2}} \qquad \text{... (3.1)}$$

By placing the values in equation 3.1 we obtain

$$D = 1.88\ d \qquad \text{... (3.2)}$$

So it is clear that, the oxide layer thickness is larger than the Si thickness that is consumed to form that oxide.

3.2 Growth and Kinetics

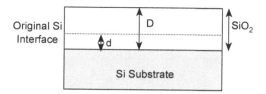

Fig. 3.3 Formation of SiO$_2$ Layer on the surface of Silicon Wafer

Thermal oxidation of silicon can be accomplished by heating the wafer to a very high temperature typically 950 to 1300°C, in an furnace containing either pure oxygen or water vapors.

3.2.1 Dry Oxidation

In dry oxidation process, the wafer is put in a pure oxygen gas (O$_2$) atmosphere. The chemical reaction takes place between the solid silicon atoms (Si) on the surface of the wafer and the approaching oxide gas.

The reaction occurring during dry oxidation at the silicon surface is

$$Si + O_2 \rightarrow SiO_2 \quad \ldots\ldots\ldots\ldots\ldots \text{DRY OXIDATION}$$

This process is performed at 1000 to 1200°C generally. To produce a very thin and stable oxide layer the process can be performed at even lower temperatures of about 900°C. For dry oxidation, figure 3.4 shows the oxide thickness as a function of oxidation time. In this process, It may be observed that the oxidation rate does not exceed ~150 nm/h. This makes it a comparatively slow process. It can be precisely controlled in order to attain a desired thickness. The oxide films resulting from this process have a good quality compared to those grown in a wet environment. This makes them more required when high quality oxides are required. Dry oxidation is normally used to produce films not thicker than 110 nm or as a second step in the growth of thicker films. The second step is utilized to improve the quality of the thick oxide.

Dry oxidation possess following characteristics:
- Slow oxide growth
- High density
- High breakdown voltage

Figure 3.4 shows empirically determined design curves. These provides time and temperature required to produce a particular layer thickness. The curve shown in figure 3.4. corresponding to dry-oxygen atmosphere.

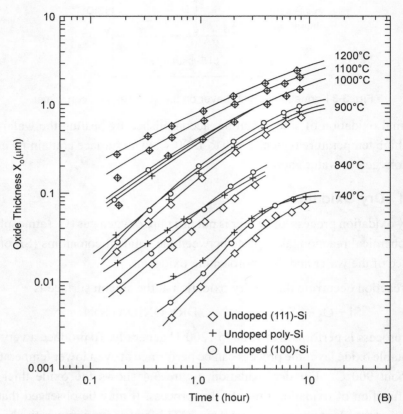

Fig. 3.4 Growth Rate of SiO_2 in a dry Oxygen Atmosphere

3.2.2 Wet Oxidation

In wet oxidation process, the silicon wafer is placed into an atmosphere of water vapor (H_2O). The chemical reaction takes place between the water vapor molecules and the solid silicon atoms (Si). In this oxidation process hydrogen gas (H_2) is released as a byproduct.

And for wet oxidation is

$$Si + 2H_2O \rightarrow SiO_2 + 2H_2 \dots\dots\dots\dots \text{WET OXIDATION}$$

This process is done by 900 to 1000°C. The wet thermal oxidation possess following characteristics:

- Very fast growth
- less quality compared to dry oxides

Figure 3.5 shows the characteristics curve for wet oxidation. It shows oxide thickness as a function of oxidation time.

3.2 Growth and Kinetics

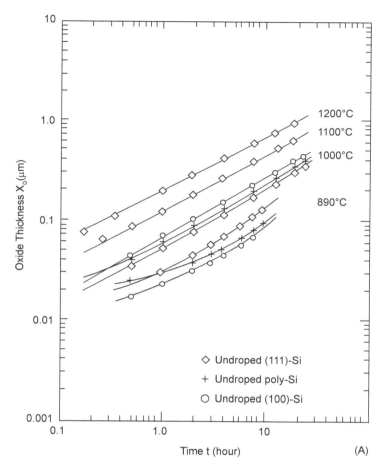

Fig. 3.5 Growth Rate of SiO_2 in a steam atmosphere

It is clear that wet oxidation process has much higher oxidation rates compared to dry oxidation, by approximately 600 nm/h. The reason for this higher oxidation rate is the ability of hydroxide (OH^-) to diffuse through the already-grown oxide much quicker than O_2. This widens the oxidation rate bottleneck when growing thick oxides. Having the fast growth rate, wet oxidation is normally used where thick oxides are required, for example masking layer, insulation and passivation layers.

Table 3.2 Comparison of growth rate of dry and wet oxidation

Wet Oxidation	Dry Oxidation
When oxidizing atmosphere contains water vapour, temperature between 900°C-1200°C	When oxidizing atmosphere contains oxygen, temperature between 900°C-1200°C
$Si + 2H_2O \rightarrow SiO_2 + 2H_2$	$Si + O_2 \rightarrow SiO_2$
Used to grow thick oxide layer called field oxide layer	Used to grow thin oxide layer

Table 3.3 Comparison of growth rate of dry and wet Oxidation with temperature

Temperature	Dry Oxidation	Wet Oxidation
900°C	18 nm/h	110 nm/h
100°C	50 nm/h	450 nm/h
110°C	130 nm/h	640 nm/h

3.3 GROWTH RATE OF SILICON OXIDE LAYER

Depending on thickness, oxides layers are grown at temperatures in the range 800–1100°C. Most oxide layers grown on Si have thicknesses more than 32 nm. But with advances in scaling, thin oxides having size of 5–20 nm are required for some applications. During advance stages of fabrication it might not be reasonable to heat the oxide to very high temperatures, because this could damage the rest of the device. At room temperature, neither the silicon nor the oxygen molecules are sufficiently mobile to diffuse through the native oxide. After a while, the reaction effectively stops and the oxide will not get much thicker than 25 Å. For a sustained reaction to occur, the silicon wafer must be heated in the presence of an oxidizing ambient.

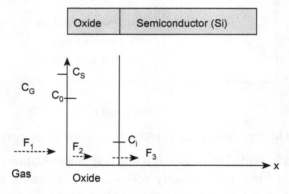

Fig. 3.6 Deal/Grove Model for thermal oxidation

Deal/Grove model is perfect model for describing the oxide growth kinetics. For both dry and wet oxidation, this model is normally effective for temperatures between 800 and 1400°C, partial pressures between 0.2 and 1.0 atmospheres, and oxide thicknesses between 0.03 and 2 μm for both wet and dry oxidation. To understand this model, consider figure 3.6, and let:

C_G = oxidant molecules concentration in bulk gas

C_s = oxidant molecules concentration adjacent to oxide surface

C_o = oxidant molecules equilibrium concentration at oxide surface

C_i = oxidant molecules concentration at Si/SiO$_2$ interface

3.3 Growth Rate of Silicon Oxide Layer

F is an oxygen flux, which represent the number of oxygen molecules that crosses a plane of a certain area in a certain time. One can then define three oxygen fluxes of interest. First of all, oxygen moves from the bulk of the gas to the surface of the growing oxide film. The oxidizing species are transported from the bulk gas to the gas/oxide interface with flux F_1 (where flux is the number of molecules crossing a unit area per unit time). The species are transported across the growing oxide toward the silicon surface with flux F_2, and react at the Si/SiO$_2$ interface with flux F_3.

The gas velocity in this boundary layer ranges from zero at the surface of the wafer to the bulk gas velocity at the opposite side of the boundary layer. To a first approximation, the oxygen molecules cannot be transported across this region by the gas flow: there is none. Instead, they must diffuse in a manner described by Fick's first law

1. Flux, F_1, is the transport of oxidizing species. Flux is transported from the bulk of the gas phase to the surface of the oxide layer, i.e. interface of gas and oxide. F_1, can be written w.r.t concentration as

$$F_1 = h_G (C_G - C_S) \quad \ldots (3.3)$$

where h_G = mass transport coefficient in the gas phase.

This can also be rewritten in terms of oxidizing species concentration within the oxide layer as

$$C_G = P_G/kT \quad \ldots (3.4)$$
$$C_S = P_s/kT \quad \ldots (3.5)$$
$$F_1 = h_G(P_G - P_s)/kT = h_G(C^* - C_0)/kT$$
as by henry law $\quad C_0 = H \times P_s \ \& \ C^* = H \times P_G \quad \ldots (3.6)$
$$F_1 = h(C^* - C_0) \quad \ldots (3.7)$$

where h can be related to h_G by

$$h = \frac{h_G}{HkT} \quad \ldots (3.8)$$

where T = temperature

H = Henry's law constant

C* = the equilibrium bulk concentration of the oxidizing species

P_s and P_G = the partial pressures of the oxidant molecules near to SiO$_2$ surface and in the bulk gas, respectively.

2. Flux, F_2, is also the transport of the oxidizing species. Here flux is transported through the oxide layer to the oxide-Si interface. For ease it can be considered that there is no dissociation of the oxidizing species within the oxide layer.

The concentration gradient needed to drive diffusion arises because the gas ambient acts as an oxygen source while the reacting surface acts as a sink. Then, assuming no sources or sinks of oxygen in the growing oxide, the concentration varies linearly and

$$F_2 \approx D_{O_2}(C_o - C_i)/d \qquad \ldots (3.9)$$

Where D_{O_2} = diffusivity of oxygen in SiO_2.

d = thickness of the oxide layer at a particular time.

3. Flux, F_3, is the reaction of the oxidizing species with Si. F_3 forms a new oxide layer. This rate is calculated by chemical reaction kinetics. Since there is an abundant supply of silicon at the reacting surface, the reaction rate and the flux are proportional to the oxygen concentration

$$F_3 = k_s C_i \qquad \ldots (3.10)$$

At steady-state, all three fluxes (F_1, F_2 and F_3) should be exactly equal. $F_1 = F_2 = F_3$ will give

$$C_i = \frac{C^*}{\left(1 + \frac{k_s}{h} + \frac{k_s \times d}{D}\right)}$$

and

$$C_0 = \frac{\left(1 + \frac{k_s \times d}{D}\right) C^*}{\left(1 + \frac{k_s}{h} + \frac{k_s \times d}{D}\right)}$$

In case, oxidation growth rate depends only on the supply of oxidant to the Si/SiO_2 interface, it is said to be "diffusion controlled. Under this condition, D is close to zero. Therefore:

$$C_i \sim 0 \text{ and } C_0 \sim C^* \qquad \ldots (3.11)$$

If, on the other hand, there is plenty of oxidant at the interface, the growth rate depends only on the reaction rate. This situation is called "reaction-controlled." In this case, D approaches infinity, and:

$$c_I = c_0 = \frac{c^*}{\left(1 + \frac{k_s}{h}\right)} \qquad (3.12)$$

Now, the growth rate can be calculated easily. Assume N_1 be the number of oxidant molecules per cm^3. After this, following differential equation can be written:

$$\frac{d}{dt} N_1(d) = F_3 = k_s \times C_i = \frac{k_s C^*}{\left(1 + \frac{k_s}{h} + \frac{k_s \times d}{D}\right)} \qquad \ldots (3.13)$$

3.3 Growth Rate of Silicon Oxide Layer

Under the boundary condition at $d = 0$ when $t = 0$, by solving the first order differential equation we get:

$$d^2 + Ad = B(t + T) \qquad ...(3.14)$$

where d = thickness of the oxide layer at time t. A, B, and T are constants.

$$A = \left(\frac{1}{k_s} + \frac{1}{h}\right)$$

$$B = \frac{2DC^*}{N_1}$$

$$T = \frac{d_i^2 + Ad_i}{B}$$

where d_i = initial oxide thickness.

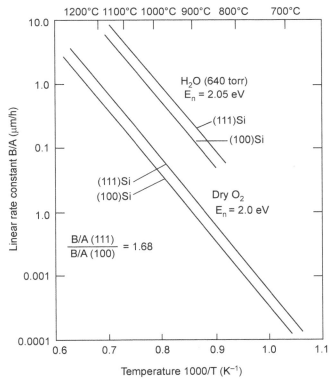

Fig. 3.7 Linear rate constants, A and B, for different types of oxide, as a temperature function. The activation energy is similar but oxidate rate is higher in wet oxidation compared to dry oxidation. (Courtesy VLSI fabrication principles – S.K. Gandhi)

The values of A and B and hence T depend on the type of oxidation (wet or dry) and also the Si surface plane i.e. (100) or (111). This comparison is shown in figure 3.7. Equation 3.14 is a second order quadratic equation. A general solution for this equation can be given as

$$\frac{d_0}{A/2} = \left(1 + \frac{t+T}{A^2/4B}\right)^{1/2} - 1 \qquad \ldots (3.15)$$

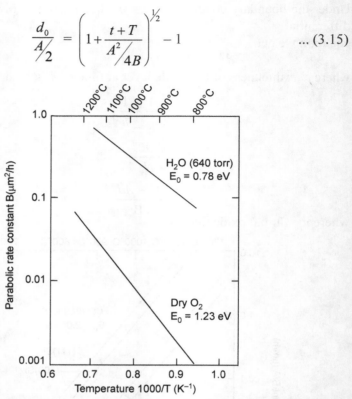

Fig. 3.8 Rate constant B for thicker oxide growth. Due to the difference in diffusion species the activation energy for wet oxidation is lower than the dry oxidation. (Courtesy VLSI fabrication principles – S.K. Gandhi)

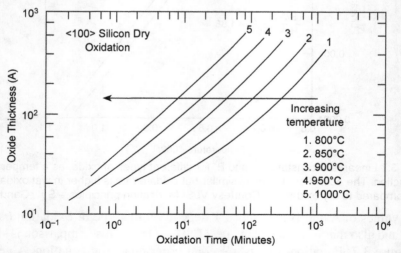

Fig. 3.9 Thin oxide growth rate for Si (100) at different temperature (Courtesy VLSI fabrication principles – S.K. Gandhi)

Effect of Temperature on Oxide Thickness

Oxidation rate can increase significantly by increasing temperature of the oxidation environment in wet as well dry processes. For dry as well wet oxidation, figure 3.4 and figure 3.5 shows temperature dependence on the oxidation rate.

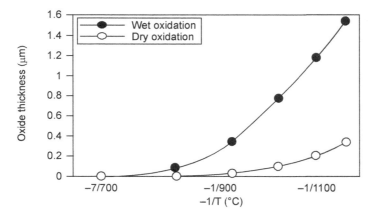

Fig. 3.10 Relationship between thickness and temperature (at 1000°C) for wet (H_2O) and dry (O_2) oxidation

Figure 3.10, shows the oxide thickness and temperature ratio. It suggests that there exists a exponential relationship between the thickness (d) and inverse negative temperature

$$d \propto e^{-1/T}$$

The diffusivity (D) of oxygen and water depends largly on temperature,

$$D \propto e^{-c/T}$$

where c = constant independent of temperature.

As, oxidant diffusivity increases exponentially with increase in temperature, so at same line oxidation rates should increase. This happens because the diffusivity of oxidants is the rate-limiting step when thicker oxides (~30 nm) are grown.

3.4 IMPURITIES EFFECT ON THE OXIDATION RATE

The oxidation rate is affected by various impurities such as :
1. Water
2. Sodium
3. Group III and V elements
4. Halogen

Besides these impurities silicon damage also affects oxidation rate. The wet oxidation has a significantly higher rate than dry oxygen, so any unintentional moisture accelerates the dry oxidation. High concentrations of sodium influence the oxidation rate by changing the bond structure in the oxide, thereby enhancing the diffusion and concentration of the oxygen molecules in the oxide. Both water vapor and oxygen diffuse easily through SiO_2 at this high temperature. This is shown in figure 3.11.

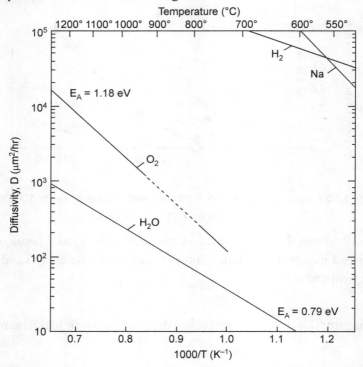

Fig. 3.11 Diffusion of Hydrogen, Oxygen, Sodium and water vapor in Silicon Glass [Copyright John Wiley & Sons Courtsey Ref. [26]

During thermal oxidation process, silicon is separated from silicon dioxide through a interface. As oxidation advances, this interface proceeds into the silicon. Silicon contains a doping impurity, that is redistributed at the interface. Across the interface, this redistribution may result in an sudden change in impurity concentration. Here the equilibrium segregation coefficient is defined as dopant in silicon to that in SiO_2 at the interface. The dopants redistribution at the interface effects the oxidation behavior. If the dopant get segregated into the oxide and remains there (for example Boron, in an oxidizing ambient), it weakens the bond structure. This weakened structure allows an increased incorporation and diffusivity of the oxidizing species. This increases the oxidation rate. Some impurities like aluminum, indium and gallium first get segregate into the oxide and later diffuse rapidly through. These impurities does not effect the oxidation kinetics. In Phosphorus impurity, impurity segregation occurs in Si rather than SiO_2. The same holds true for other impurities like As and Sb dopants.

Oxidation rate cam be enhanced by using Halogen (such as chlorine) impurities. It is introduced intentionally. Following points plays a role in increasing oxidation rate

1. By reducing sodium ion contamination,
2. By increasing oxide breakdown strength,
3. By reducing interface trap density.

Traps can be considered as levels in the forbidden energy gap. These are linked with defects in the silicon.

3.5 OXIDE PROPERTIES

When oxidation layer is grown in a dry atmosphere, it results higher density. Higher density means low impurities and a better oxide quality compared to oxide growth in wet atmosphere.

Oxide's volume may expand or shrink when it is exposed to range of various temperature, and is represented by Thermal expansion. This parameter is very low, for oxides meaning it does not put ample stress and strain on other materials those are in contact with it.

Oxide's stiffness can be measured with Young's modulus and Poisson's ratio calculates negative ratio of oxide's transverse to axial strain. These two are important measures of a material's mechanical stability.

The thermal conductivity, affects power during operation. It has been also proved that the thermal conductivity of oxides changes depending on the oxide thickness. The typical value for thermal conductivity are

Oxide Type	Thermal conductivity
Thin sputtered oxide	1.1 W/m-K
Thin thermally grown oxide	1.3 W/m-K
Bulk oxide	1.4 W/m-K

The high dielectric strength shows the stability of SiO_2 under high electric fields. It suggests that the oxide film is very suitable for dielectric isolation.

Strong, directional covalent bonds, forms the silicon dioxide. SiO_2 possess a well local structure having four oxygen atoms. These atoms are arrayed at the corners of a tetrahedron around a central silicon atom. The tetrahedra cells is shown in Figure 3.12a.

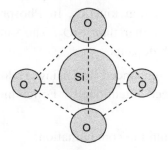

Fig. 3.12(a) Si-O bond structure

The tetrahedral bond together by sharing oxygen atoms as illustrated in Figure 3.12b in a sample four membered ring.

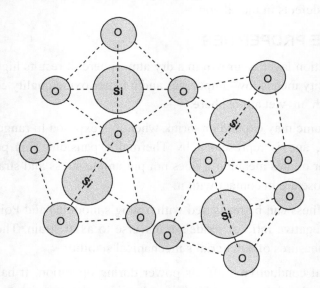

Fig. 3.12(b) Four-membered ring

3.6 OXIDE CHARGES

The interlace among Si and SiO_2 contains a transition region. Generally numerous charges are related with the oxidized silicon, some of which are associated to the transition region. A charge produced at the interface can induce a charge of the opposite polarity in the underlying silicon. This affects the ideal features of the MOS device. This results in both reliability and yield problems. The various types of charges are shown in figure 3.13.

3.6 Oxide Charges

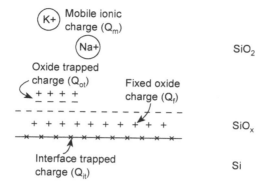

Fig. 3.13 Different charges occurred in thermally oxidized silicon

Interface-trapped charges

This type of charge, is the charge due to electronic energy levels located at the Si-SiO$_2$ interface with energy states in the silicon band gap that can capture or emit electrons (or holes). These electronic states arise because of the lattice mismatch at the interface, dangling (incomplete) bonds, the adsorption of foreign impurity atoms at the silicon surface, and several other defects produced by bond-breaking processes or radiation processes. These are the most important type of charges because of their wide-ranging and degrading effect on device behavior.

Fixed oxide charge

It is located very close to the Si-SiO$_2$ interface. It is generally positive. This charge is located in the oxide within approximately 30Å of the Si-SiO$_2$ interface. Fixed oxide charge cannot be charged or discharged; its density is not greatly affected by the oxide thickness or by the type or concentration of impurities in the silicon, but it depends on oxidation and annealing conditions, and on silicon surface orientation.

Mobile ionic charge

Alkali ions like sodium, potassium, and lithium present in the heavy metals are responsible for this charge. At room temperature in presence of electric field, alkali ions are mobile. To prevent mobile ionic charge contamination of the oxide during device life, one can protect it with a film impervious to mobile ions such as amorphous or small-crystallite silicon nitride. For amorphous Si$_3$N$_4$, there is very little sodium penetration. Other sodium barrier layers include Al$_2$O$_3$ and phosphosilicate glass.

Oxide trapped charge

The oxide trapped charge is linked with faults in SiO_2, and may result from avalanche injection or ionizing radiation. The oxide traps are usually electrically neutral and are charged by introducing electrons and holes into the oxide through ionizing radiation such as implanted ions, X-rays, electron beams, etc. The magnitude of trapped charge depends on the amount of radiation dose and energy and the field across the oxide during irradiation.

3.7 OXIDATION TECHNIQUES

Figure 3.14 shows various methods to grow oxide later. These methods are discussed in detail in Chapter 8.

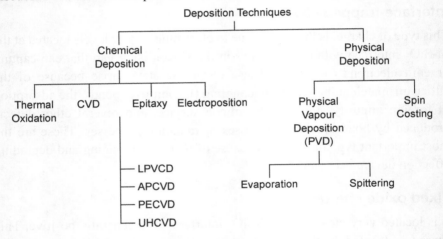

Fig. 3.14 Oxide layer deposition techniques

3.8 OXIDE THICKNESS MEASUREMENT

The oxide thickness is an important parameter of the oxidation process, and thus many ways have been developed to measure it. Here we will describe few of the several methods for estimating the thickness of an oxide. Each has its inherent advantages and disadvantages. Every method make some assumption about the oxide that may be valid only under certain circumstances. Aside from the techniques described here there are a number of thin film measurement techniques that can be used.

1. Physical determination of the oxide thickness requires the production of a step in the oxide. Typically this is done with a mask followed by an etch. Hydrofluoric acid (HF) etches oxide at a much higher rate than it etches silicon. Therefore, if a mask is applied to the wafer, the wafer

3.8 Oxide thickness measurement

immersed in HF, and then the mask is removed, a step nearly equal to the oxide thickness will be left. This step can be measured using a scanning electron microscope (SEM) if it is larger than 200 Å, or with a transmission electron microscope (TEM) if it is not.

2. An easier approach is to use a surface profilometer, an instrument that measures surface topology by mechanically scanning a needle stylus while it is in contact with the wafer. The deflection of the needle is measured, amplified, and displayed as a function of position. Resolution of these instruments down to 2 Å is claimed by the manufacturers. Similarly, atomic force microscopy (AFM) can be used to measure the step. Profilometry has the advantage that it makes no assumptions other than the relative etch rates of the oxide and the silicon. Since part of the oxide must be etched to determine the thickness, this test is destructive and generally requires the use of a dedicated test wafer.

3. The simplest optical technique is to partially immerse the unmasked wafer in dilute HF until the oxide on the submerged portion of the wafer is completely removed. Near the line between the etched and unetched oxide, a slow grading of the thickness will be found. If this edge is examined under a microscope a variety of colors will be seen starting from light brown (Table 3.4). These colors are due to interference between the incident and reflected light. By following the colors up to the top of the oxide an approximate thickness can be found.

Table 3.4: Color Chart for silicon dioxide (refractive index of 1.48)

Colour	SiO_2 Thickness (Å)	Si_3N_4 Thickness (Å)
Silver	< 270	< 200
Brown	< 530	< 400
Yellow-brown	< 730	< 550
Red	< 970	< 730
Deep blue	< 1000	< 770
Blue	< 1200	< 930
Pale Blue	< 1300	< 1000
Very Pale Blue	< 1500	< 1100
Silver	< 1600	< 1200
Light yellow	< 1700	< 1300
Yellow	< 2000	< 1500
Orange-red	< 2400	< 1800
Red	< 2500	< 1900
Dark red	< 2800	< 2100
Blue	< 3100	< 2300

Colour	SiO₂ Thickness (Å)	Si₃N₄ Thickness (Å)
Blue-green	< 3300	< 2500
Light green	< 3700	< 2800
Orange-yellow	< 4000	< 3000
Red	< 4400	< 3300

Note that multiple orders exist. An SiO_2 film that appears red may be 730 to 970 Å, 2400 to 2500 Å, or 4000 Å.

4. Ellipsometry technique uses, a polarized coherent beam of light that is reflected off the oxide surface at some angle. Helium–Neon lasers are commonly used as a source.

   The reflected light intensity is measured as a function of the polarization angle. To measure refraction index and film thickness, reflected and incident intensity are compared and change in polarization angle is measured. To do this definitively requires measurement at more than one incidence angle or for more than one wavelength as more than one thickness generates the same change in the light at any given angle or wavelength.

   Variable angle spectroscopic ellipsometers systematically vary both angle and wavelength and fit the data to a model to extract thickness and index. Ellipsometry has the advantage of being nondestructive, although it often requires that the oxide be grown on bare silicon. Since the ellipsometer's beam is quite large, it is also normally done on unpatterned wafers. It is also common to measure many points across the wafer and map the film thickness.

5. Electrical techniques are the most useful way to characterize oxide layers. The simplest electrical measurement is the breakdown voltage. The voltage on the capacitor is increased while the electrical current through the oxide film is measured. The leakage current is too small to measure until a high electric field is reached. Eventually, for thin oxides, a current will be detected that will rise exponentially with voltage. Within a small voltage range the current increases discontinuously, signaling an irreversible rupture of the oxide. The dielectric field strength of thermal oxide is about 12 MV/cm. Thus, by knowing the breakdown voltage, one can estimate the oxide thickness. If, as is usually the case, the thickness is known, the breakdown field can be measured. A breakdown histogram is often presented as a first-order indication of oxide quality and defect density. Three groups of breakdown regions are evident. The low voltage group is known as extrinsic breakdown. These are killing defects such as pinholes in the growth process. The Breakdown field (MV/cm) high voltage

group is intrinsic breakdown. They typically cluster near the ultimate breakdown field of the oxide. The intermediate group is usually associated with weak spots in the oxide. The larger the fraction of breakdown events contained in the intrinsic break-down group, the better the oxide quality. To be significant, the area of the capacitor tested should be similar to the active area on a chip.

6. Capacitance-Voltage (C-V) measurements are more sensitive methods of evaluating oxides. Again, a metal film must be used as an upper electrode. The wafer is used as a lower electrode. Assume for the moment that the substrate is doped p-type. On application of negative voltage to the gate, additional holes are drawn to the Si–SiO$_2$ interface. This process is known as accumulation. Now assume that a small ac signal is added to the dc bias and the ac current is measured. The out-of-phase magnitude of current is proportional to the capacitance. The device will be considered as parallel plate capacitor, if the diameter of the capacitor is very large compared to the oxide thickness. Then

$$C_{ox} = \frac{\varepsilon A}{t_{ox}} = \frac{\varepsilon_r \varepsilon_0 A}{t_{ox}} \qquad \text{... (3.16)}$$

where ε_r is the relative permittivity or dielectric constant of the oxide and ε_o is the permittivity of free space. Again, this is a valuable method of measuring the oxide thickness. For very thin oxides however, one must also take into account the finite width of the accumulation layer in the semiconductor. If the bias voltage is ramped from the negative set point to positive values, the measured capacitance will decrease. This occurs because the field changes sign, repelling the charge directly below the gate. This effectively increases the width of the dielectric, reducing the capacitance.

3.9 OXIDE FURNACES

Batch process is used to grow thermal oxides in tube furnaces, i.e. manifold wafers are processed at the same time. In the context of process control, this becomes important as any change from required conditions would affect all wafers in that batch and hence lead to overall cost increase. The furnace is designed to have a long flat zone in which the temperature can be controlled from 400°C to 1200°C within ±1°C. One end of the furnace has provisions for the flow of pure dry oxygen or water vapour while the other end opens into a vertical flow clean air bench where the wafers can be loaded into the reactor. The hood is designed to keep out particulate matter and minimize contamination during wafer loading. Gas flow, insertion and withdrawal of wafers as well as the furnace temperature are micro-processor controlled. The furnace temperature is

ramped up and down several times to prevent thermal shock to the wafers and damage later. Utmost cleanliness is essential in wafer handling as well as in maintenance of the diffusion tube which must be cleaned at intervals. In special cases the slotted quartz boat can be replaced by one made of polysilicon. For wafer having small sizes, typically 3" and 4" wafers, horizontal tube furnaces are utilized for oxidation. This furnace is shown in figure 3.15. The furnace has three zones:
1. Source zone
2. Center zone
3. Load zone.

The gases required for oxidation are introduces through source zone. Generally, oxygen (for dry oxidation) and steam (for wet oxidation) at the appropriate partial pressure (concentration) is used as a source gas. In some applications, chlorinated oxide layers are also grown. The chlorine incorporated in the oxygen gives various advantages. It reduces charge concentration at the oxide-Si interface and mobile ions in the oxide layer.

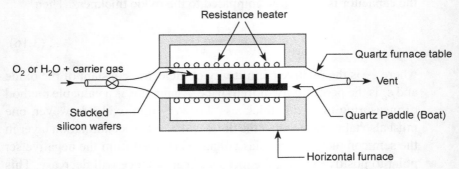

Fig. 3.15 Horizontal Diffusion Furnace (Courtsey: Microchip fabrication)

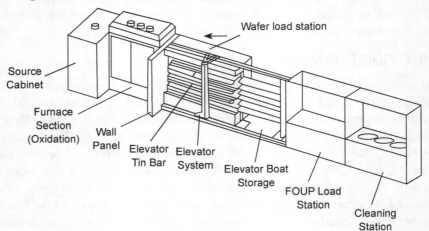

Fig. 3.16 Commercial Horizontal Diffusion Furnace
(Courtsey Microchip fabrication - Peter van Zant)

3.9 Oxide furnaces

This results in cleanliness and improved device performance. Chlorine is introduced in the furnace in the form of Cl_2, hydrogen chloride gas or trichloroethylene. Vapor from gaseous sources is mixed with the oxygen, while for liquid sources, the gas is bubbled through the liquid.

Before introduction of oxygen in the furnace, some steps are taken to reduce contamination. These steps are known as few purge and pump steps. The commercially available tube furnaces have also three zones:

1. Wafer loading zones
2. Cleaning stations zone
3. Wafer storing zone

A commercial horizontal furnace is shown in figure 3.16. The wafers to be processed are loaded in the center zone. Generally, baffle plates of quartz are loaded at the both ends. In addition to process wafers, bare wafers, known as fillers, are also loaded. These fillers aids in regulating gas flow through the furnace. With this uniform oxide growth of process wafers is obtained. In should be kept in mind that, all wafers in the furnace are not process wafers. Higher process throughput can be obtained by maintaining high ratio of process wafers to blank wafers. Process throughput can be defined as the number of process wafers processed per hour. During oxidation process, temperatures need to be constantly monitored, maintained and regulated within the furnace. A PID (proportional-integral-derivative) mechanism is used for this temperature control.

If wafer is of large size (common wafers are now 300 mm), horizontal furnaces are not practical. These occupies lot of space. For large size wafers vertical diffusion furnaces are used. Vertical furnace is also known as diffusion furnace. The VDF is shown in figure 3.17. The furnace consists of:

1. Wafer loading station
2. Wafer storing station (before and after processing).

The boat that keeps the wafer to be processed moves vertically into the furnace. VDF offer two major advantages, first it is compact than horizontal furnaces, gas flow is more uniform and less turbulence. The boat is rotated continuously during operation to maintain uniformity of the oxide layer. This is specifically true for mixed gases. Mixed gases move parallel to gravity and hence do not get separated. The VDF operation is similar to the horizontal tube furnace. Typically, a 150 wafer boat holds a maximum of 100 product wafers, the rest are fillers, baffles, and monitor wafers (for measuring oxide thickness and uniformity for process control).

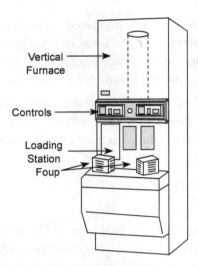

Fig. 3.17 Vertical Diffusion Furnace (Courtsey Microchip fabrication-Peter van Zant)

3.10 SUMMARY

The two factors that allowed silicon to dominant semiconductor material in use today are 1) SiO_2 is a native oxide that provides a high quality insulating barrier on the surface of silicon wafer 2) SiO_2 can serve as a barrier layer during subsequent impurity diffusion process steps. The basic growth mechanism is oxidant transport through the SiO_2 layer to the Si/SiO_2 interface where a simple chemical reaction produces the new layers of oxide. This chapter introduced the topic of the thermal oxidation of silicon, presenting the Deal–Grove model. This model accurately predicts the oxide thickness of a wide range of oxidation parameters. Enhanced growth rates are seen for thin oxides. Effect of impurity on oxide, oxide charges, oxide quality and oxide thickness measurement techniques are also described as well. As the understanding of oxide charge is necessary in order to fabricate highly reliable devices. Oxidation thickness can be accurately measured using ellisometers, interference microscopes and mechanical surface profiles or can be estimated from the apparent color of the oxide under vertical illumination with white light. Finally, typical oxidation systems are described.

PROBLEMS

1. Which one is better between dry oxidation and wet oxidation and why?
2. Why initial oxidation is linear and it becomes parabolic thereafter?

3. How are the interface charges measured in the presence of fixed charges by the capacitance voltage measurement technique?
4. A silicon wafer of N type, (100), 10^{-2} Ω-cm is oxidized at 1100°C for 220 minutes in wet oxidation. Calculate the oxide thickness.
5. Calculate the gate oxide thickness of a MOS transistor for a 1.2 volt threshold voltage with surface charge Q.
6. Why is steam oxidation more rapid than dry O_2 oxidation?
7. Under What conditions is the thermal growth rate of SiO_2 linearly proportional to time?

REFERENCE

1. D.A. Buchanan and S.H. Lo, "Growth, in The Physics and Chemistry of SiO_2 and the Si-SiO_2 Interface-3," 3-14 *The Electrochemical Society*, Pennington (1996).
2. L.C. Feldman, E.P. Gusev and Garfunkel, "Fundamental Aspects of Ultrathin Dielectrics on Si-based Devices," 1-24, *Kluwer Academic Publishers*, Dordrecht, (1998).
3. http://www.siliconfareast.com/$SiO_2Si_3N_4$.htm.
4. Bearbeitet von and Hamid Bentarzi, "The MOS structure, Transport in Metal-Oxide-Semiconductor Structures," *Engineering Materials, DOI*: 10.1007/978-3-642-16304-3_2.
5. M.A. Muhsien, I.R. Agool, A.M. Abaas and K.N. Abdalla, "Current transport in SiO_2 films grown by thermal Oxidation for metal-oxide semiconductor," *International Research Journal of Engineering Science, Technology and Innovation* (IRJESTI), 1(2):25 (2012).
6. http://www.purdue.edu/rem/rs/sem.htm
7. http://en.wikipedia.org/wiki/Scanning_electron_microscope
8. K. Schroder "Semiconductor Material and Device Characterization", Third Edition, Dieter, Arizona State University Tempe, AZ, IEEE press, John Wiley & Sons, (2006).
9. Lee Stauffer, "Fundamentals of Semiconductor C-V Measurements," Keithley Instruments, Inc.
10. H.U. Kim and S.W. Rhee, "Electrical Properties of Bulk Silicon Dioxide and SiO_2/Si Interface Formed by Tetraethylorthosilicate-Ozone Chemical Vapor Deposition," *Journal of The Electrochemical Society*, 147 (4):1473 (2000).
11. http://web1.caryacademy.org/facultywebs/gray_rushin/StudentProjects/CompoundWebSites/2003/silicondioxide

12. M. Liu, J. Peng, "Two-dimensional modeling of the self-limiting oxidation in silicon and tungsten nanowires". T*heoretical and Applied Mechanics Letters*. 6 (5):195. doi:10.1016/j.taml.2016.08.002 (2016).

13. http://www.eng.tau.ac.il/~yosish/courses/vlsi1/I-4-1-Oxidation.pdf

14. J. Appels, E. Kooi, M.M. Paffen, J.J.H. Schatorje, and W.H.C.G. Verkuylen, "Local oxidation of silicon and its application in semiconductor- device technology," *Philips Research Reports*, 25(2):118 (1970).

15. A. Kuiper, M. Willemsen, J.M.G. Bax, and F.H.P.H. Habraken, "Oxidation behaviour of LPCVD silicon oxynitride films," *Applied Surface Science,* 33(34), 757 (1988).

16. D.A. Buchanan and S.-H. Lo, "Growth, Characterization and the Limits of Ultrathin SiO_2 Based Dielectrics for Future CMOS Applications," T*he Physics and Chemistry of SiO_2 and the Si-SiO_2 Interface—III*, Electrochemical Society, Pennington, NJ, 3 (1996).

17. H.S. Kim, S.A. Campbell, D.C. Gilmer, and D.L. Polla, "Leakage Current and Electrical Breakdown in TiO_2 Deposited on Silicon by Metallorganic Chemical Vapor Deposition," *Appl. Phys. Lett.* 69:3860 (1996).

18. S.A. Campbell, D.C. Gilmer, X. Wang, M.T. Hsieh, H.S. Kim, W.L. Gladfelter, and J.H. Yan, "MOSFET Transistors Fabricated with High Permittivity TiO_2 Dielectrics," *IEEE Trans. Electron Dev.*, 44:104 (1997).

19. K.J. Hubbard and D.G. Schlom, "Thermodynamic Stability of Binary Oxides in Contact with Silicon, *in Epitaxial Oxide Thin Films II,*" 401, Pittsburgh, 33 (1996).

20. T. Ino, Y. Kamimuta, M. Suzuki, M. Koyama, and A. Nishiyama, "Dielectric Constant Behavior of Hf-O-N system," *Jpn. J. Appl. Phys.*, 45(4 B): 29082913 (2006).

21. S.A. Campbell, T.Z. Ma, R. Smith, W.L. Gladfelter and F. Chen, "High Mobility HO_2 N-and P-Channel Transistors," *Microelectron. Eng.* 59(1-4):361 (2001).

22. Z. Zhang, B. Xia, W.L. Gladfelter, and S.A. Campbell, "The Deposition of Hafnium Oxide from Hf t-butoxide and Nitric Oxide," *J. Vacuum Sci. Technol.* A, 24(3):418-423 (2006).

23. J.H. Lee, Y.S. Suh, H. Lazar, R. Jha, J. Gurganus, Y. Lin, and V. Misra, "Compatibility of Dual Metal Gate Electrodes with High-p Dielectrics for CMOS," *IEDM, Tech. Dig*, 323-326 (2003).

24. F. Chen, B. Xia, C. Hella, X. Shi, W.L. Gladfelter, and S.A. Campbell, "A Study of Mixtures of HO_2 and TiO_2 as High-k Gate Dielectrics," *Microelectron. Eng.* 72(1-4):263-266 (2004).

25. I. McCarthy, M.P. Agustin, S. Shamuilia, S. Stemmer, V.V. Afanas'ev, and S.A. Campbell, "Strontium Hafnate Films Deposited by Physical Vapor Deposition," *Thin Solid Films*, 515(4):2527 (2006).

26. S.K. Gandhi, "VLSI Fabrication Principle," John Wiley and Sons, New York (1983).

27. J. Appels, E. Kooi, M.M. Paffen, J.J.H. Schatorje and W.H.C.G. Verkuylen, "Local oxidation of silicon and its application in semiconductor- device technology," *Philips Research Reports*, 25(2):118 (1970).

28. A. Kuiper, M. Willemsen, J.M.G. Bax, and F.H.P.H. Habraken, "Oxidation behaviour of LPCVD silicon oxynitride films," *Applied Surface Science*, 33(34):757 (1988).

29. J.D. Plummber, M. Deal and P.B. Griffin, "Silicon VLSI Technology: Fundamentals, Practice and Modelling," Pearson (2009).

30 S.M. Sze, "VLSI Technology," Second Edition, McGraw Hill Education (India) Private Limited, (1988).

31. S.K. Gandhi, "VLSI Fabrication Principles," Second Edition Wiley, (1994).

32. S. Grove, "Physics and Technology of Semiconductor," John Wiley & Sons. (1967).

33. S.A. Campbell, "The Science and Engineering of Microelectronic Fabrication," Oxford University Press, (1966).

4

Lithography

4.1 INTRODUCTION

Patterning the functional material is crucial for all technologies from ancient era to present DNA microarrays world. Advantages of pattering such as expansion to shrinking capability, higher speed due to reusable capability, precision in reproducibility and lower energy consumption per computing function makes it crucible for complex geometry of ICs. Continuous miniaturization of devices takes IC into nanometer level so producing the same design in several places uses pattering process. Patterning on the wafer in most ICs utilizes two steps: a) patterning of a resist film on top of the functional material, process is known as lithography and b) transferring the resist pattern into the functional material, by the process called as etching.

Lithography is the most complicated, expensive, and critical process in mainstream microelectronic fabrication. The term lithography curtails from the art world made by impressing, in turn, several flat reproduced slabs, each covered with ink of some colour, on a paper or canvas. The various layers must be accurately aligned to each other within some registration tolerance. Thus, many prototypes can be made from the same set of slabs, as long as the quality remains high enough. This is the basic principle in IC lithography. The process allows mass production of components that are almost identical to within required accuracy. The circuit pattern is directly written on or projected to the wafer or resist with the aid of a mask. This mask can be used multiple times to produce near identical components.

Lithography accounts for about one-third of the total fabrication cost, a percentage that is rising. A typical silicon technology will involve 15–20 different masks. In IC technology, Lithography is the process of transferring pattern of design outline through mask to the surface of semiconductor wafer which is covered through radiation-sensitive material known as resist.

Figure 4.1(a) is graphical representation of different lithographic processes employed in IC fabrication. There are various methods that can be used to project the circuit pattern on to the wafer. In this chapter we will discuss popular lithography methods within the industry and some of the methods under development such as optical, X-Ray, electron beam, Extreme Ultra-Violet, and ion-beam lithography. Optics is most widely used technique and has continued to be the preferred lithographic route despite its continually forecast demise. A combination of 193-nm radiation, computer-intensive resolution enhancement technology and immersion optics is used for the 45- and 32-nm nodes. Optical lithography process typically requires first formation of a mask and then project that mask pattern onto a resist-coated wafer. Forming a precise and qualified mask represents a significant fraction of the total patterning cost of an IC largely because of the measures needed to push resolution so far beyond the normal limit of optical resolution. Thus, although optics has demonstrated features well below 22 nm but it is not clear that optics will be the most economical in this range. Extreme ultraviolet is still the official front runner then nanometer-scale mechanical printing is a strong contender with electron beam lithography that has demonstrated minimum features less than 10 nm wide and continuous developing both for mask making and for directly writing on the wafer (also known as maskless lithography). Figure 4.1(b),

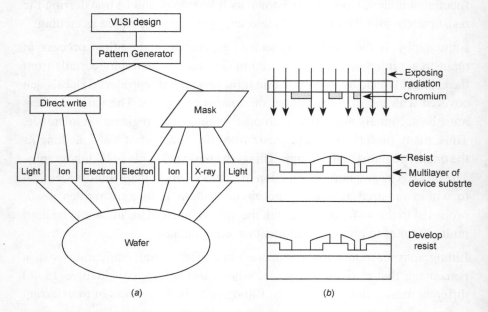

Fig. 4.1 (a) Lithographic Process (b) Optical Lethographic Process

shows the general optical lithography process in which the radiation is transmitted through the clear parts of the mask which exposed photoresist insoluble in the developer solution thus enabling the direct transfer of the mask pattern onto the wafer. An etching process is hired as the patterns are defined on the wafer for selectively remove masked portions of the underlying layer.

The lithographic exposure performance is determined by three parameters: (a) resolution (b) registration (c) throughput. Resolution is the minimum feature dimension that can be transferred with high reliability to a resist film on wafer. Registration is a measure of how precisely patterns on sequential masks can be aligned or overlaid with respect to previously defined patterns on the same wafer. Throughput is the number of wafers that can be exposed per hour for a given mask level so it is a measure of the efficiency of the lithographic process.

4.2 OPTICAL LITHOGRAPHY

Optical equipment using ultraviolet light ($\lambda \cong 0.2$ μm to 0.4 μm) or deep ultraviolet light is popular lithographic equipment for IC fabrication. There are mainly two optical exposure methods: (1) shadow printing (2) projection printing. In shadow printing, the mask and wafer may be in direct contact known as contact printing, or in close proximity called proximity printing (Figure 4.2) while in projection printing, an image of the mask patterns have projected through exposure tool onto a resist-coated wafer many centimeters away from the mask.

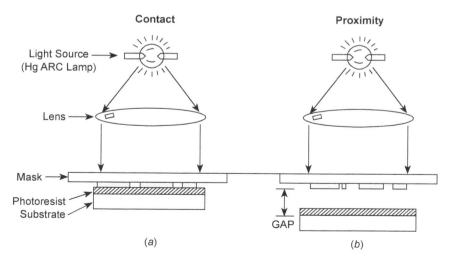

Fig. 4.2 Optical Lithography

4.3 CONTACT OPTICAL LITHOGRAPHY

The first integrated circuits in the early 1960s were patterned through contact lithography that places a mask directly on top of the resist. The process of making masks has following steps:

(*i*) Fabricate an enlarged outline (reticle) by cutting a thin plastic sheet of Rubylith which choked blue light.

(*ii*) Expose the reticle pattern at compact magnification onto a master mask blank that contained a quartz or glass substrate coated with 80-nm chromium and above that a photosensitive polymer (resist).

(*iii*) Repeat the exposure throughout entire master mask to build up an array of identical patterns on the mask. The instrument was referred to as a 'step and repeat camera')

(*iv*) Develop the resist then etch the chromium.

(*v*) Duplicate the master mask onto silver halide emulsion plates known as daughter masks or working plates.

The daughter masks were used to contact print the required pattern on top of the wafer with resist film after that underlying functional layer was then etched into the required pattern.

In contact lithography the diffraction that may occur between mask and resist is removed by bring the mask into contact with the resist and the resolution of the system is increased without changing the exposing wavelength or the numerical aperture of the system. The mask is pressed with a pressure of typically around 0.3 atm against the resist. The system is then exposed to UV light of around 400 nm wavelength. The actual nature of contact between surfaces means that resolution will vary across the resist but resolution of around 0.5 µm or less is possible. The contact results in distortion of both the resist and the mask so the mask can only be used for a short time compared to proximity lithography process. However, the masks are relatively cheap to produce and with the deformity to the resist are acceptably small; contact lithography is a better method to use because it produces an improved resolution. It can overcome the issue of non-uniform resolution across the resist by introducing flexible masks that help to obtain better contact with the resist material to get a more uniform resolution. Typical materials include PMMA (Poly-methyl methacrylate), Al_2O_3 and Quartz.

4.4 PROXIMITY OPTICAL LITHOGRAPHY

Proximity optical lithography requires no image formation between the mask and the resist. The proximity system is basically made up of a light source,

4.5 Projection Optical Lithography

a condenser, a mirror, a shutter, a filter here known as mask and the stage on to which the resist is positioned as illustrated in figure 4.3. In proximity lithography, the mask-resist separation is usually around 5 to 50 µm that leads to an acceptable resolution for today's devices of around 200 nm. However, diffraction occurs between the mask and the resist that affects resolution. Narrowing of the mask- resist separation required for better resolution however a much better resolution would be achieved if the resist and the mask were in contact.

Contact printing suffers from major drawback caused by dust particles or silicon specks accidentally embedded into the mask so causing permanent damage to the mask while proximity printing is not affected by particle damage. However, the narrow gap between the mask and wafer introduces optical diffraction at the feature edges on the photo masks that degraded and the resolution typically to the 1- 3 µm regime.

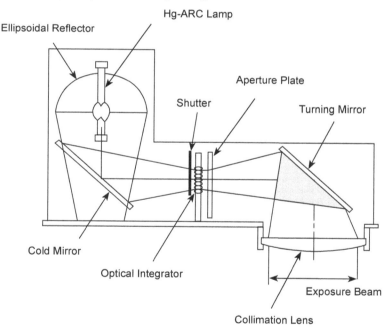

Fig. 4.3 Proximity optical Lithography

4.5 PROJECTION OPTICAL LITHOGRAPHY

Projection optical lithography offers highest resolution(less then 0.35µm) than proximity and contact printing. Projection printing relies on an image formation system between the mask and the resist because the beam is focused between the mask and the resist wafer. The complete process is illustrated in figure 4.4. The projection system consists of several sub-systems, each of which can be manipulated to improve the overall resolution of the system. The numerical apertures of the lenses can be increased up to around 1.6 that is the practical limit since sin α term has a maximum of 1 and η typically 1.5 to 1.7. A greater improvement in resolution will be given by decreasing the wavelength of the exposing radiation so wavelengths of Ultra-Violet (UV) and X-ray utilizes in lithographical systems. However sufficient improvements to optical lithography can be made to ensure it remains the VLSI production technique as phase-shifting masks are under development within the industry as well as lasers are being used within the industry to give more coherent light sources. VLSI circuits that have a dimension of 180 nm for each section of the circuit pattern uses partially coherent light from a KrF laser ($\lambda = 248$ nm) in combination with some resolution enhancement techniques including phase-shifting masks and advanced resists. Lasers with yet shorter wavelengths are being developed for further enhance the resolution utilizing ArF ($\lambda = 193$ nm) and F_2 ($\lambda = 157$ nm). It has been recorded that the best resolution for $\lambda = 193$ nm is 150 nm, while for $\lambda = 157$ nm is 125 nm.

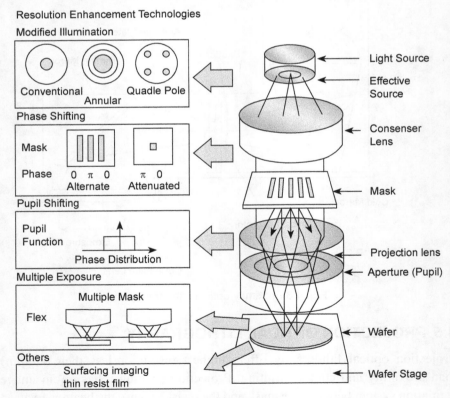

Fig. 4.4 Projection optical Lithography

4.5 Projection Optical Lithography

The IC industry is constantly endeavoring for reductions in component sizes. Therefore, at some point in the future a resolution better than 125 nm will be required. To achieve this, the lithographic industry either has to improve optical lithography by using phase-shifting masks (discussed later in this chapter) as well as look to shorter wavelength radiation.

Diffraction, Resolution and Depth of Focus

One of measuring factors in optical lithography is resolution because diffraction occurs due to the mask that essentially acts as a set of slits within the optical lithography system. The type of diffraction in the lithographic systems is determined first. As the waves from the light source are emitted spherically so also get spherical waves from the mask. If the resist film is close to the mask then the curvature of the spherical wave is important. This result in an image which will be similar to the mask and in near field Fresnel diffraction occurs. As the distance between the mask and the resists increases the image will look less like the mask with a little change in the shape of the image beyond a certain distance, known as the Rayleigh distance (R_D), only its size will change. This is due to the curvature of the spherical wave becoming negligible, and in this far field diffraction occurs known as Fraunhofer diffraction. The Rayleigh distance, R_D is as:

$$R_D = \frac{a^2}{\lambda} \quad \ldots (4.1)$$

For explaining these consider waves coming from two different masks. The waves diffract, producing Airy's rings on the resist surface, as shown in figure 4.5 (a). If we decrease the spacing of the gaps on the mask then the Airy's rings come closer together. For the integrated circuit pattern produced by this process to be useful condition is the rings must be distinguishable, or resolvable as shown in figure 4.5 (b).

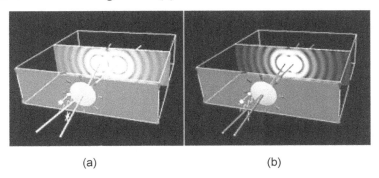

(a)   (b)

Fig. 4.5 Airy's Rings resulting from diffraction of a parallel beam of light passing through a pair of mask.

Now, consider the distance between two of these spots must be in order to be distinguishable. The two images are distinguishable when the maximum intensity of one set of Airy's Rings overlaps with the first minima of the second set of Airy's Rings as shown in figure 4.6.

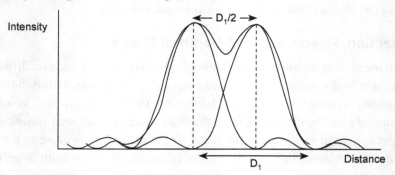

Fig. 4.6 The intensity of Airy's Rings from two adjacent masks. The combined profile is presented as a dotted line. This figure shows the Rayleigh resolution limit, equal to a distance of $D_1/2$ on the resist

Figure 4.6 show that a distance of $D_1/2$ must separate the peaks of intensity in order to determine the two points. The apertures subtend a semi-angle α at the resist surface. Now we can derive an equation for the minimum resolvable separation ($D_1/2$) by manipulating the basic properties of the system, for that we consider light diffracting from a single circular aperture, as shown in figure 4.7

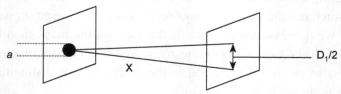

Fig. 4.7 Diffraction from a single circular mask having diameter a. The intensity maximum (P) and the first minimum (P') are separated by $D_1/2$ on a resist at a distance X from the mask

From basic optics, we have that

$$\frac{D_1}{2} = 1.22 \frac{X\lambda}{a} \qquad \ldots (4.2)$$

Now consider the case of projection optical lithography, where a lens is placed between the mask and resist. The distance X becomes the focal length (f) of the lens. So equation 4.2 becomes

$$\frac{D_1}{2} = 1.22 \frac{f\lambda}{a} \qquad \ldots (4.3)$$

4.5 Projection Optical Lithography

The width of the mask can be eliminated from equation 4.3 by considering the angle it subtends from the resist, as shown in figure 4.8

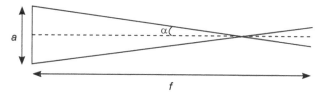

Fig. 4.8 The mask subtends a semi-angle α at the resist. It is used to replace the width of the mask with it's semi-angle in a mathematical model for resolution.

From figure 4.8, an equation for the width of the mask in terms of its distance from the resist and its semi-angle can be derived as

$$\sin \alpha = \frac{a}{2f} \qquad \ldots (4.4)$$

by replacing a in equation 4.3, to give us an equation for minimum resolvable separation in terms of the focal length of the lens and the semi-angle of the aperture

$$\frac{D_1}{2} = \frac{0.61 \lambda}{f \sin \alpha} \qquad \ldots (4.5)$$

Noted thing at this point is that the focal length has refractive index dependence. If wrap up the other constant features of the focal length with the 0.61 factor, the equation can be re-written as

$$R = \frac{k_1 \lambda}{\mu \sin \alpha} \qquad \ldots (4.6)$$

Here, R is the minimum resolvable separation ($D_1/2$), k_1 is a constant dependent on the absorbing properties of the resist and the type of radiation used in the process, λ is the wavelength of the exposing radiation and η is the refractive index of the resist. Numerical Aperture (NA) is usually referred by the quantity $\eta \sin \alpha$. Therefore, equation 4.6 becomes

$$R = \frac{k_1 \lambda}{NA} \qquad \ldots (4.7)$$

Now consider the depth of focus of the process. The image formed in an optical system is only brought to focus in the appropriate plane i.e. a sphere. The depth of field is the distance over which the image retains an acceptable focus, as shown in figure 4.9.

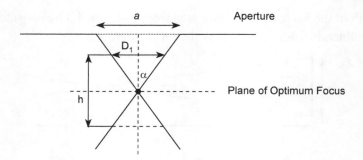

Fig. 4.9 The depth of focus of an optical system, represented here as h, is the distance either side of the plane of optimum focus over which the image retains an acceptable focus.

The depth of field is related to D_1 as with resolution. By simple geometry as:

$$h = \frac{k_2 \lambda}{NA^2} \quad \ldots (4.8)$$

Considering equations 4.7 and 4.8, it is found that resolution and depth of field are in conflict. Depth of field is improved by increasing the wavelength of the radiation and by decreasing the numerical aperture. As resolution improves on decreasing the wavelength and increasing the numerical aperture so a compromise is needed between resolution and depth of focus. In present equipment, the positioning of the resist within the lithographical process can be controlled accurately so depth of field becomes less of an issue. Noted that as the wavelength decreases the photon energy increases so the photons will able to penetrate further in the resist polymer. As the depth of field becomes minor factor so the industry can concentrate on improving resolution. Obviously, the industry cannot neglect the depth of focus of the process, but resolution become the primary factor.

4.6 MASKS

IC fabrication is done by mass batch processing, where many copies of the same circuit are deposited on a single wafer and many wafers at the same time. The number of wafers processed at one time is called the Lot whose size may vary between 20 to 200 wafers. Since each IC chip is square and the wafer is circular, the number of chips per wafer is the number of complete squares of a given size that can fit inside a circle.

The photographic mask controls the location of all windows in the oxide layer and so areas over which a particular diffusion step is effective. Each complete

4.6 Masks

mask consists of a photographic plate on which each window is represented by a dense part and remainder remains transparent. Each complete mask will not only include all the windows for the production of one stage of a particular IC, but in addition, all similar areas for all such circuits on the entire silicon wafer.

It will be obvious that a different mask is required for each stage in the production of an array of IC's on a wafer. There is also a vibrant requirement for precise alignment between one mask and the other in series to ensure that there is no overlap between components and that each section of a particular transistor is formed in precisely the correct location.

To make a mask of the production stages a master is first prepared that is an exact replica of that portion of the final mask associated with one individual integrated circuit, but which is certain time enlargement of the final size of IC. Art work at enlarge size avoids large tolerance errors and also permits to be dealt easily by human operator. In the design of the art work the locations is all

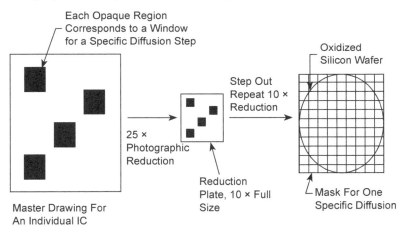

Producing one of a series of photographic masks required for the manufacture of an array of ICs. Figures are not a scales.

Fig. 4.10 Mask Production

components such as resistor, capacitor, diode, transistors and so on are determined on the surface of the chip. Therefore six or more layout drawings are required. Each drawing shows the position of windows that are required for a particular step of the fabrication. For VLSI ($> 10^5$ components/chip), the patterns are generated using computer-aided design (CAD) systems whose output drives a pattern generator that transfers the patterns directly to the photosensitive masks. Masks are typically made from fused silica i.e. glass covered with hard-surface materials such as Cr or Fe_2O_3. The circuit pattern is first transferred to the electron resist which is transferred underlying layer for the finished mask.

4.7 PHOTOMASK FABRICATION

The master mask is typically of order 1 m x 1 m, prepared from cut and strip plastic material which consists of two plastic films, one photographically opaque called Rubilith and the other transparent called mylar are laminated together. The outline of the pattern required is cut in the red coating of Rubilith (that is opaque) using a machine controlled cutter on an illuminated drafting table. The opaque film is then peeled off to disclose transparent areas thus each part represents a window region in die final mask.

The next step is to photograph the master using back illumination, to produce a few times reduced sub-master plate. This plate is used in a step and repeat camera which serves two purposes first reducing the pattern by a further few times to finished size and second is also capable of being stepped mechanically to produce an array of identical patterns on the final master mask. Better accuracy may be achieved by using continuous plate movement instead of the photographic plate being transported mechanically in discrete steps. Discrete exposures being made by an electronically synchronized flash lamp which effectively freezes the motion.

The entire sequence just described can be done with plates containing a photosensitive emulsion; typically the emulsion is considered too weak to abrasion and tears. For this reason, masks are often made of harder materials such as Cr or Fe_2O_3. The composite layout is divided into several sequences so 15-20 diverse mask levels are required to complete IC fabrication cycle.

One of the major concerns of mask is defect density. Mask defects can be presented during the manufacturing of the mask or during succeeding lithographic processes. The number of mask defects has a deep effect on the final IC yield that is defined as the ratio of good chips per wafer to the total number of chips per wafer. First-order approximation as:

$$Y \cong e^{-D_0 A_0} \qquad \ldots (4.9)$$

where Y is the yield, D_0 is the average number of "fatal" defects per unit area and A_0 is the defect sensitive area of an IC chip. If D_0 remains the same for all mask levels (e.g., N =10 levels), then

$$Y \cong e^{-N D_0 A_0} \qquad \ldots (4.10)$$

Figure 4.11 illustrate the mask-limited yield for a 10-level lithographic process as a function of chip size for various values of defect densities.

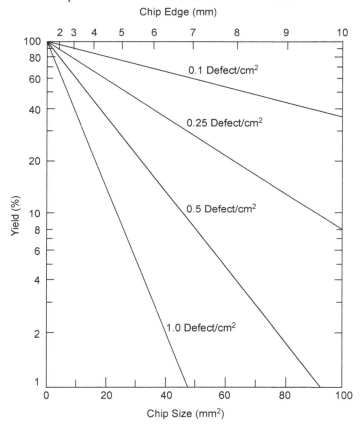

Fig. 4.11 Mask Yield

4.8 PHASE SHIFTING MASK

A phase-shifting mask is employed to diminish the problems arising from wavelength or depth of focus (DOF). The basic concept of the phase-shifting mask is illustrated in figure 4.12. At the conventional transmission mask (Figure 4.12a), the electric field ξ has the same phase at every transparent area. The limited resolution and diffraction of the optical system spread the electric field ξ at the wafer as intensity I is proportional to the square of the electric field shown by the line. Interference between waves diffracted by the consecutive spots in mask enhances the field between them.

The phase-shifting layer that covers adjacent spot in mask reverses the sign of the electric field but intensity at the mask is unchanged as shown in figure 4.12(b). The electric field of these images shown by the dotted line can be canceled at the wafer. Subsequently, images that are projected close to each other can be separated completely. A 180° phase change produces when a transparent layer of thickness d = $\lambda / 2$ ($\eta - 1$), where η is the refraction index and λ is the wavelength, covers one mask as shown in figure 4.12(b).

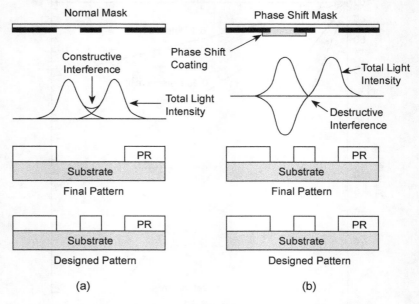

Fig. 4.12 Basic concept of Phase Shifting Mask

4.9 PHOTORESIST

To get an authentic recording of the geometry of mask over the substrate surface the resist should fulfill following conditions:

- Uniform film formation
- Resolution
- Good adhesion to the substrate
- Proper resistance to dry and wet etch processes

A photoresist is a radiation-sensitive compound that forms a Polymer film in radiation. The film is photosensitive or capable or reacting with the photolysis product of added compound so that the solubility in developer solution increases or decreases significantly by exposure to UV radiation. According to the solubility changes that take place, photoresists are termed negative or positive. Materials which are solidified, less soluble in a developer solution by radiance produce a negative pattern of the mask and are called negative photoresists. Wherein positive resists, the exposed region becomes more

4.9 Photoresist

soluble by radiance so more readily removed in the developing process. The net result is that the patterns formed in the positive resist on the wafer are same as those on the mask and the patterns etched are the reverse of the mask patterns for negative resists as the exposed regions become less soluble. The change in solubility is due to a set of chemical reactions which occur as the photon scatters and loses energy in the resist polymer material. Figure 4.13 shows lithographic transfer process.

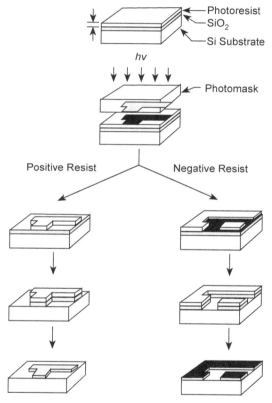

Fig. 4.13 Lithographic transfer process

A positive photoresist comprises of three ingredients: (a) photosensitive compound, (b) base resin, (c) organic solvent. The photosensitive compound is insoluble in the developer solution earlier to UV radiance. During radiation the photosensitive compound in the exposed pattern areas changes its chemical structure by absorbing the radiation energy (free energy) and converts into a more soluble species. Bonds are then broken within the resin itself that increases its solubility. After that the exposed areas are removed by the developer solution. In positive resists exposed section of the resist is not required for the final integrated circuit pattern. This is known as shadow printing. Few positive electron resists are PMMA, and poly-butene-sulfone also known as PBS.

Negative photoresists are polymers pooled with a photosensitive compound i.e. it consist of a chemically inert film-forming component along with a photoactive agent. During exposure the photosensitive compound absorbs the free energy and converts it into chemical energy for initiating a chain reaction in which photoactive agent releases nitrogen gas on exposure to light and the radicals generated in this reaction react with the C=C and C=O double bonds within the polymer that causes crosslinking of the polymer molecules. The cross-linked polymer becomes hard due to higher molecular weight so insoluble in the developer solution. The unexposed portions are removed after development processing. During this process, the cross-linked polymer molecules tend to swell as they now have a higher molecular weight, and therefore distort the pattern on the resist. The major drawback of a negative photoresist is limiting the resolution as the resist absorbs developer solvent and swells. Common negative electron resists is Poly-glycidyl-methacrylate-co-ethyl-acrylate abbreviated COP.

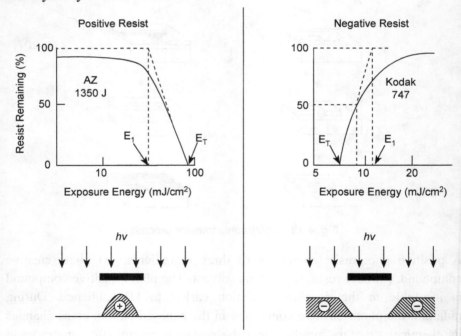

Fig. 4.14 Positive photoresist (left) and negative photoresist (right): Exposure response curve and cross section of the resist image after development

The left portion of figure 4.14 shows the exposure response curve for a positive resist. It should be noted that even prior to exposure the resist has a finite solubility in the developer solution. At a threshold energy (E_T) the resist becomes completely soluble so E_T corresponds to the sensitivity of the photoresist. The contrast ratio (γ) has dependent on E_T as given:

$$\gamma = \left[\ln \frac{E_T}{E_1} \right]^{-1} \qquad \ldots (4.11)$$

Where E_1 is the energy obtained by sketching the tangent at E_T to reach 100% resist thickness as shown in figure 4.14. A larger γ indicates rapid dissolution of the resist with an incremental rise of exposure energy that resulted a sharper image. The image cross section depicted in (figure 4.13) illustrates that the edges of the resist image are generally blurred due to diffraction.

The right portion of Figure 4.14 illustrate an analogous situation but for a negative photoresist. Here the sensitivity of a negative photoresist is defined as the energy required retaining 50% of the original resist film thickness in the exposed region. Table 4.1 lists some of the common resists used in IC Technology.

Table 4.1 Common resists used in IC Technology.

Lithography	Name	Type	Sensitivity	γ
Optical	Kodak 747	Negative	9 mJ/cm²	1.9
	AZ-1350J	Positive	90 mJ/cm²	1.4
	PR102	Positive	140 mJ/cm²	1.9
e-beam	COP	Negative	0.3 µC/cm²	0.45
	GeSe	Negative	80 µC/cm²	3.5
	PBS	Positive	1 µC/cm²	0.35
	PMMA	Positive	50 µC/cm²	1.0
X-ray	COP	Negative	175 mJ/cm²	0.45
	DCOPA	Negative	10 mJ/cm²	0.65
	PBS	Positive	95 mJ/cm²	0.50
	PMMA	Positive	1000 mJ/cm²	1.0

4.10 PATTERN TRANSFER

The goal of pattern transfer is to transfer the sketches on the mask to the wafer surface. To achieve this, two phases are often performed when fabricating microelectronic devices. In the first phase, a photolithography process is used to transfer the pattern of the mask on the photoresist; in the second phase, the thin-film removal is employed to copy the image on the photoresist on to the wafer surface. In what follows, we are describing the first phase in more detail. The details of thin-film removal are deferred to the next chapter.

The major steps required for pattern transfer in the photolithography process are shown in figure 4.15. In the following, we take a look at these major steps along with some minor enhancement steps.

Wafer clean and prime

The first step of the photolithography is to clean the wafer because it might be contaminated during the previous step. Then, a priming process is utilized to deposit a thin primer layer to wet the wafer surface as shown in figure 4.15(a) that enhance the adhesion between the photoresist and the water surface.

Photoresist coating

The wafer is coated with a liquid photo resist by a spin coating method, as shown in figure 4.15(b). The spin speed and the viscosity of photoresist material determine the final photoresist thickness, ranging from 0.6 to 1 μm.

Softbake

Before going to the alignment and exposure step, a softbake process is required to drive off most of the solvent in the photoresist material. The softbake process is to place the wafer on a hot plate at a temperature of 90 to 100°C for about 30 seconds.

Alignment & Exposure

The mask is aligned to the correct location of wafer coated with the photoresist and then exposed into a controlled UV light to transfer the mask image onto the photoresist surface as displayed in figure 4.15(c)

Post-exposure bake

The post-exposure bake is intended to minimize striations of overexposed and underexposed areas through the photoresist caused by the standing-wave effect that might be occurring from the interference between the incident light and the light reflected from the photoresist-substrate interface. In modern processes, a thin Anti Reflective Coating (ARC) layer is often used to help reduce the amount of reflective light.

Development

Development is the critical step for creating the pattern in photo resist on the wafer surface. In this step, the soluble regions are removed by developer chemicals, as shown in figure 4.15 (d). After development, the following two steps are often carried out.

Hardbake

After development, the wafer needs to be baked again on a hot plate at a temperature of 100 to 130°C for about 1 to 2 minutes to drive out the remaining solvent in the photoresist material. This step improves not only the strength and adhesion but also the, etch and ion-implantation resistance of the photoresist.

Photoresist Stripping

The remaining part of the pattern transfer of the running example is completed by etching away the silicon dioxide exposed and then removing the photoresist. Succeeding oxide etching and with the help of abrasion process, the remaining resist is finally stripped off with a mixture of H_2SO_4 and H_2O_2. Finally a step of washing and drying completes the required window in the oxide layer. The figure below shows the silicon wafer ready for next diffusion. The illustrations of these two steps are depicted in figures 4.15 (e) and (f), respectively.

Pattern inspection

The closing step of the pattern transfer process is pattern inspection, which checks whether the sketch on the mask correctly transported onto the photoresist. The photoresist has to stripped and the whole process is repeated again if the wafer fails in inspection test.

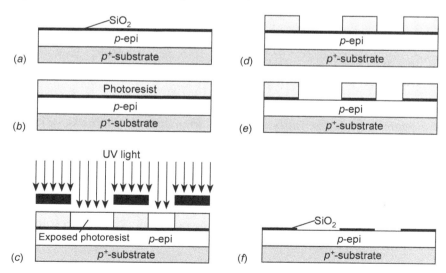

Fig. 4.15 Pattern Transfer process

The insulator image can be engaged as a mask for subsequent processing. For example, ion implantation can be executed to dope the exposed regions selectively. Figure 4.16 illustrates the liftoff technique which serve if the film thickness is smaller than that of the photoresist.

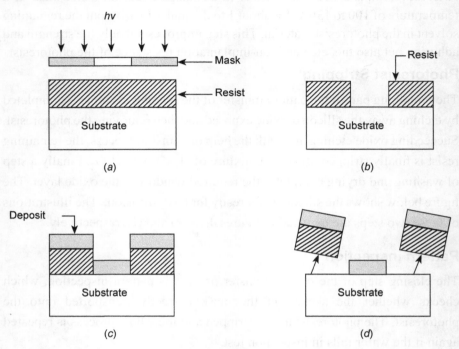

Fig. 4.16 Liftoff Process

4.11 PARTICLE-BASED LITHOGRAPHY

4.11.1 Electron Beam Lithography

Tinier resolutions are requirement by the IC industry that achieve through the use of shorter wavelengths. Based on the absorbing capacity of mask or resist material Gamma rays wavelength cannot be used. Utilizing the de Broglie equation (equation 4.12) for principle of wave-particle duality an electron having energy of 10 keV produces a wavelength of around 12 pm. That signifies a huge reduction in wavelength compared to X-ray radiation. This implies that Electron beam lithography has the possibility of better resolution than any of the electromagnetic methods.

$$\lambda = \frac{h}{\sqrt{2me\Delta V}} \qquad \ldots (4.12)$$

Electron beam lithography replaces the photons with an electron beam, and utilizes a different optical system for image formation known as direct writing

4.11 Particle-Based Lithography

between the source and the resist. Electron beam lithography process as shown in figure 4.17.

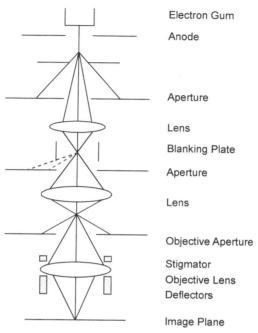

Fig. 4.17 Electron beam lithography system.

Electron beam lithography offers better resolution because of the small wavelength of electrons (< 0.01 nm for 10-50 keV electrons). The resolution of an electron lithographic system is limited electron scattering in the resist (Figure 4.18) and by the various aberrations of the electron optics not by diffraction. The main advantages of electron beam lithography are listed as:

- Generation up to nano and submicron resist geometries
- Precisely controlled and highly automated operation
- Superior depth of focus
- Direct writing i.e. patterning without a mask

The use of electrons whose direction can be controlled by a magnetic field provides no requirements to practice a mask in the lithographic system. A computer controls the strength of the magnetic field so the patterns produced on the resist are extremely accurate even though chance of scattering in the resist and little diffraction from the electrons. Truly, the patterns are so accurate that electron beam lithography is used to construct the masks for the electromagnetic radiation techniques.

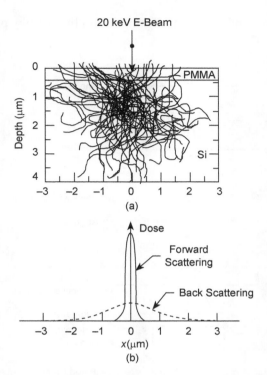

Fig. 4.18 (a) Simulated trajectories of 100 electrons in PMMA for 20 keV electron beams (b) Dose distribution for forward scattering and back scattering at the resist-substrate interface.

Electron beam lithography offers a more accurate technique with higher resolution but there are some problems associated, as the system has a very low throughput, and it is slow process. Electron beam lithography is a slow process because of writing very fine patterns on the resist and only one point on the resist can be exposed at any given time. In accordance to Moore's Law, the patterns have become dense regularly electron beam lithography also has an association in mask manufacture so the fabrication of the masks has become increasingly slow and this has become unaffordable to the industry. Therefore, concentrating on shaping and enlarging the size of the beam can improve the throughput in electron beam lithography. For example, the capability to varying electron beam shape and size means that the writing process is independent of the minimum feature size. However decreasing the time to complete the process need to develop systems that are capable of parallel exposure.

4.11.2 Electron-Matter Interaction

Electrons considered as waves do not interact with matter in the same way as electromagnetic waves in the lithographic system. In fact when electrons interact with matter they are considered as charged particles and not waves.

4.11 Particle-Based Lithography

There occurs the resistance of electron exposure by breaking of bonds or by the formation of bonds between polymer chains. The incident electrons have far greater energies than the bond energies of the molecules in the resistance that's why there occurs a molecular reactions exposing beam of electrons. Exposing beam of electrons and bond formation (due to the electrons being captured) and bond breaking (due to the energies of the electrons) occur simultaneously, the dominant process being determined by the nature of the resist (positive or negative) as shown in figure 4.19. There is the dominancy of bond formation and the electron-induced crosslinks between molecules make the resist polymer less soluble in the developer solution. Indeed one crosslink per molecule is more than enough to make the polymer insoluble, therefore allowing us to remove the remainder of the resist, leaving just the pattern made from the insoluble polymer behind. If larger molecules are used, then we need less crosslinks per unit volume to make the polymer insoluble. In a positive resist, the bond breaking process subjugates. Exposure to the electron beam results in a lessening of molecular weight since molecules in the polymer are broken up. This leads to greater solubility than the unexposed parts of the resist, allowing us to remove the exposed part and leave the rest of the resist material. The major factor that limits electron beam lithography is scattering of electrons. When electrons are incident on the resist, they enter the material and lose energy by a series of collisions this process is known as electron scattering, and produces secondary electrons, X-rays, heat and some electrons may also be back scattered. This characteristic of electrons interacting with matter limits the resolution of the system.

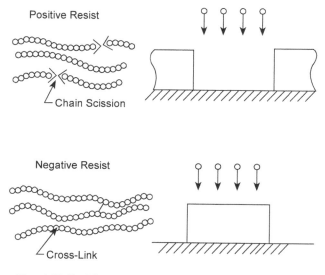

Fig. 4.19 End formation in positive and negative resist.

Resists

There is a formation of bonds between polymer chains when negative resistance is exposed to electron beam and bond breaking occurs in positive resist when it is exposed. The electron-beam induced cross-links between molecules of negative resist make the polymer less soluble in the developer solution. Resist sensitivity increases with increase in molecular weight. In positive resistance the bond breaking process predominates thus exposure leads to lower molecular weight and greater solubility. The polymer molecules in the unexposed resist will have a distribution of length or molecular weight and thus a distribution of sensitivities to radiation. The narrower the distribution, the higher will be the contrast. High molecular weight and narrow distribution are advantageous. There is also a fundamental process limitation on resolution, when electrons are incident on a resist or other material, they inter the material and lose energy by scattering, and produces secondary electrons and X-rays. This limits the resolution to an extent that depends on resist thickness, beam energy, and substrate composition. Resolution is better for thinner layer of resist. while use of device processing minimum thickness is set in order to keep the defect density low and resistance of etching. For photo masks where the surface is flat and only a thin layer of chrome must be etched with a liquid etchant, resist thickness in the range of 0.2 to 0.4 μm are used. In case of more severe dry gas plasma etching process employed, thickness of 0.5 to 2 μm are required. One way to overcome this problem is to use a multilayer resist structure in which the thick bottom layer consists of the process-resistant polymer. A three-layer resist structure can be used in which the uppermost layer is for patterning a thin intermediate layer, such as SiO_2 which serves as a mask for etching the thick polymer below. For electron lithography a conducting layer can be substituted for the SiO_2 layer to prevent charge build-up that can lead to beam placement errors. Multilayer resist structure also improves the problem of proximity effect encountered during electron beam exposure. An exposed pattern element adjacent to another element receives exposure not only from the incident electron beam but also from scattered electrons from the adjacent elements. A two-layer resist structure is also used in such structure, both the thin upper and the thick lower layer are positive electron resist, but they are developed in different solvents. The thick layer can be overdeveloped to provide the undercut profile that is ideal for liftoff process.

Electron Optics

The first extensive use of electron-beam pattern generators in photo mask making is discussed in previous section. The electron-beam exposure system (EBES) machine has proved to be the best photo mask pattern generator.

Scanning electron-beam pattern generators are similar to scanning electron microscopes. A basic probe-forming electron optical system consists of two or more magnetic lenses and provisions for scanning the image and blanking the beam on the wafer image plane. There is a typical image spot size in the range of 0.1 to 2 µm which is far from the diffraction limits therefore diffraction can be ignored. However, aberrations of the final lens of the deflection system will increase the size of the spot and can change its shape as well.

Electron Projection Printing

Electron projection system provides high resolution over a large field with high throughput. Rather than a small beam writing the pattern in serial fashion, a large beam provides parallel exposure of large area pattern. In a 1:1 projection system parallel electric and magnetic fields image electrons onto the wafer. The mask is of quartz and is patterned with chrome. It is covered with ICs on the side facing the wafer. Photoelectrons are generated on the mask/cathode by backside UV illumination. There is an advantage of stable mask projection system, good resolution, fast step-repeat exposure with low sensitivity electron resists, large field, and fast alignment. The limitations of the system include proximity effects of electrons and shorter life of cathode.

Electron Proximity Printing

This is a repetitive system in which a silicon membrane stencil mask containing one chip pattern is shadow printed onto the wafer. There is no accommodation of re-entrant geometries in mask. Registration is accomplished by reference to alignment mask on each chip. It has an advantage to measure and compensate mask distortion. Proximity effects must be treated by changing the size of pattern elements. The main limitation of the system is that it needs two masks for each pattern.

4.12 ION BEAM LITHOGRAPHY

In electron beam lithography we obtain an element of back scattering and removing or reducing of this factor improves resolution of the system. So there is an advantage of using ions instead of electrons because ions are more massive. From the de Broglie equation (equation 4.12), it has been noticed that massive ions have shorter wavelength which also helps in improving resolution. This fact should also be considered that the resists are more sensitive to ions than electrons. Increase in the energy of ions improves the yield to a higher extent but when the ions starts penetrating too deeply into the resist it becomes a problem. So there should be enough energy in the system to produce ions. There are two types of Ion Projection lithography (IPL) system: focused beam system and mask-beam system.

The material used in Ion Projection Lithography is Hydrogen or Helium gas. The H^+, H^{2+}, H^{3+} or He ions are extracted from a source at around 10 keV, and sent through a set of initial lenses then the ions are passed through a patterned stencil (a silicon membrane), and entered into a multi-electrode electrostatic lens system which projects a magnified image of the mask on to the resist. Whilst passing through the electrostatic lens system, the ions are accelerated up to 200 keV to ensure absorption in the resist. This technique along with several others is in the very early stages of development. IPL could sooner or later yield a resolution of the order of a nanometer with present resolutions around 5 nm.

Ion-beam lithography provides higher resolution when exposed to resist than that possible with an electron-beam because of less scattering. Resists are more sensitive to ions than to electrons a unique feature of ion-beam is that there is the possibility of wafer processing without resists if it is used to implant or sputter selected areas of the wafer. Repair of photo mask is the most important application of Ion beam lithography. Ion-lithography employs a scanning focused-beam or a masked-beam.

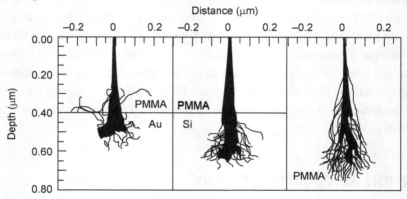

Fig. 4.20 Computer Trajectory

Alternative Lithography Using Electromagnetic Radiation

Ion optics has more severe problems than that of electron optics in order to scan ion beams. The source of ionized material is a gas surrounding a pointed tungsten tip or a liquid metal that flows to the tip from a reservoir that's why electrostatic lenses are used for focusing ion beam. If a magnetic lens were used, the field would have been much larger than in the electron optics. Normally electrostatic optical systems have higher aberrations needing small aperture and small scan fields. Figure 4.20 is depicting the computer trajectory of 50 H^+ ions implanted at 60 keV illustrating the spread of the ion beam at a depth of 0.4 µm are only 0.1 µm. There is also the possibility of resistless wafer process however, an ion beam is usually larger than an electron beam and the resolution is thus adversely affected. Repair of masks for optical or X-ray lithography is the most important application of ion lithography and is also used for commercial process.

4.13 ULTRA VIOLET LITHOGRAPHY

When wavelength is shortened from optical to UV offers a viable alternative to optical lithography as the requirement for smaller circuit patterns increases beyond the present capabilities of optical lithography. Extreme Ultra-Violet (EUV) lithography uses the same principle as the projection optical system. Rayleigh's equation still holds the expression for the depth of focus using UV wavelengths of 1-13 nm gives a resolution which is significantly better than the optical system. However; it has its own problem.

The absorption of radiation at this short wavelength is very strong therefore lens-based refractive optics cannot be used in lithographic system instead of that a reflective optical system can be used. However, conventional mirror surfaces cannot be used at these wavelengths as they are transparent, so multi-layer devices which depend on interference for reflection must be used. In the UV region, these devices usually have a reflectivity of only 60-70%, and hence the number of mirrors in the system must be kept to a minimum or the intensity level reaching the resist will be insufficient. The requirement for fewer optical components means that asymmetric mirrors need to be introduced and they should be extremely precise with errors of the order of 0.1 nm. Therefore there is a need of an extremely precise mirror manufacturing technique to be developed first. The multi-layer reflectors consist of a large number of alternating layers of materials having dissimilar optical properties which provide a constant reflectivity when the thickness of each layer is $\lambda/2$. The nature of these reflectors make manufacturing difficult, but without them EUVL would be impossible. However, the system gives a resolution of less than 100 nm.

The source of the UV radiation is another problem where radiation of this wavelength is more energetic than visible light, and therefore we must have higher-powered energy sources to produce the UV radiation. The best candidate at present is laser-produced plasma of xenon gas, although several other sources, including synchrotrons, are under expansion.

EUV faces problems with the resists therefore resists which strongly absorb EUV radiation need to be developed before the technique so that it can be fully utilized. Resists which are in development consist of fewer layers (to enable miniaturization), which also strongly absorb EUV radiation. As the features printed on the resist have become smaller, the rough edges of the printed lines become a problem. The problem stems from diffraction, as well as using normal transverse waves to try and print straight lines. This is a problem for all lithographic techniques, but a successful EUV resist would be required to solve this problem in the case of EUV.

4.14 X-RAY LITHOGRAPHY

In X-ray lithography an X-ray source illuminates a mask which casts a shadow on the resist. Overlaying the pattern, made out of X-ray absorbing material, on to a transmitting material, produces the mask used in X-ray lithography. Any materials to be used in the absorbing part of the mask must have high absorption coefficients in the X-ray region. Figure 4.21 explain the comparison between X-ray proximity lithography contrasted with optical lithography.

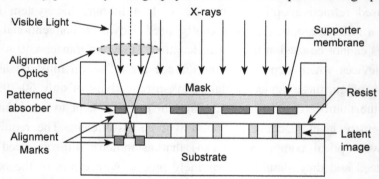

Fig. 4.21 Typical mask structure in X-ray proximity lithography contrasted with optical lithography (left hand side of figure). Before the X-rays reach the mask, they are collimated by a silicon carbide mirror and passed through a Beryllium glass window. We note that the patterns on the X-ray lithography masks, as well as the optical lithography masks, are produced by electron beam lithography

The absorption coefficient of any elementary material of atomic number Z and density ρ is proportional top $Z^4\lambda^3$ over a wide range of wavelengths. However, the proportionality constant drops in a step function fashion at the absorption edge. This is a wavelength which corresponds to the ionization energies of inner K-shell electrons X-ray lithography has a higher throughput when compared to E-beam lithography because parallel exposure can be adopted.

4.14 X-Ray Lithography

X-Ray Resist

An electron resist can also be referred to as an X-ray resist because an X-ray resist is exposed largely by the photoelectrons produced during X-ray absorption. The energies of these photoelectrons are much smaller than the 10 keV to 50 keV energies which are used in electron lithography making proximity effects negligible in the case of X-ray and promising higher ultimate resolution.

Most of the polymer resists containing only H, C, and O, absorb very small X-ray flux. This small absorption has the benefit of providing uniform exposure throughout the resist thickness and the disadvantage of reduced sensitivity.

Electron beam resists can be used in X-ray lithography because when an X-ray photon impinges on the specimen, electron emission results. One of the most attractive X-ray resist is DCOPA (dichloropropyl acrylate and glyciedyl methacrylate-co-ethyl acrylate), as it has a relatively low threshold ($\sim$ 10 mj cm^2).

Proximity Printing

Since the wavelength of X-ray is small diffraction effects can be ignored and simple geometrical considerations can be used in relating the image to the pattern on the mask. The opaque parts of the mask cast shadows on to the wafer below. The edge of the shadow is not absolutely sharp because of the finite diameter of the focal spot of the electrons on the anode X-ray source at a finite distance from the mask. However, on account of the finite size of the X-ray source and the finite mask-to-wafer gap, a penumbral effect results which degrades the resolution at the edge of a feature as shown in figure 4.22, the penumbral blur, δ, on the edge of the resist image is given by $\delta = ag / L$

Where a is the diameter of the X-ray source, g is the gap spacing, and L is the distance from the source to the X-ray mask If a = 3 mm, g = 40 µm, and L = 50 cm, δ is on the order of 0.2 µm.

An additional geometric effect is the lateral magnification error due to the finite mask-to-wafer gap and the non-vertical incidence of the X-ray beam. The projected images of the mask are shifted laterally by an amount d, called run out $d = rg / L$

Where r denotes the radial distance from the center of the wafer or a 125 mm wafer, the run out error can be as large as 5 µm for g = 40 µm and L = 50 cm. This run out error must be compensated for during the mask making process.

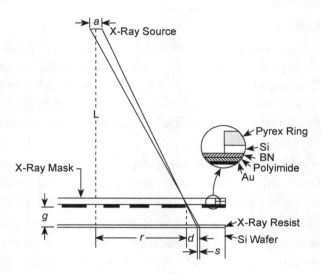

Fig. 4.22 Proximity Printing

X-Ray Sources

In earlier years of development X-ray sources were often an electron beam evaporator with its chamber modified to accept a mask and wafer. The target metal could be changed easily to modify the X-ray spectrum. X-ray generation by electron bombardment is a very inefficient process most of the input power is converted into heat in the target. The X-ray flux is generally limited by the heat dissipation in the target. Much high X-ray fluxes are available from generators which have high speed targets. Another type of source, which provides still greater amount of flux, is the plasma discharge source in which the plasma is heated to a temperature high enough to produce X-radiation. The plasma chamber has problems such as reliability and contaminations.

X-Ray Masks

The mask for X-ray lithography consists of an absorber on a Tran's missive membrane substrate. The absorber is usually gold which a heavy metal and also it can be easily patterned. The transmissive membrane substrate is a polymer such as polymide and polyethylene terephthalate.

4.15 COMPARISON OF LITHOGRAPHIC TECHNIQUES

Optical lithography is the main stream technology and some commercially available resists can resolve down to 0.1µm or lower. Usually optical lithography is considered difficult to use for a design rule of much less than 1µm due to its resolution limit. For deep sub-micrometer structures the two remaining options are electron beam direct writing or X-ray lithography. Yet perfect X-ray masks are difficult to make and the throughput of electron lithography is slow (the throughput varies as the reciprocal of the square of the minimum feature length that is $\propto \Delta l_m$). For mass production, the cost and footprint (required floor area) of the machine must also be minimized.

Table 4.2 Comparison of various Lithography techniques.

Parameter	193 nm optical	126 nm optical	UV	X-Ray	Electron Beam	Ion Beam
Effective Wavelength	193 nm	126 nm	11 nm	0.1 nm	12 pm	0.1 pm
Exposing Particles	Photons	Photons	Photons	Photons	Electrons	Ions (Photons)
Type of Optics	Transmission	Reflective	Reflective	Reflective	Electromagnetic	Electromagnetic
Mask Type	Transmission	Reflective	Reflective	Transmission	None	Transmission
Resolution Limit	100 nm	100 nm	45 nm	30 nm	22 nm	2 nm
Typical D.O.F.	400 nm	500 nm	1100 nm	Large	Large (Scattering)	Large (Scattering)

4.16 SUMMARY

The continuous growth of IC Industry is a direct result of the capability to transfer smaller and smaller circuit patterns onto semiconductor wafers.

This chapter is concentrated on the production of the areal image, the optical intensity as a function of position on the surface of the wafer for the small features of interest in integrated circuit production, diffraction effects are extremely important. Simple contact printers can be used for pattern structures to less than 1μm, but these systems are highly defecting prone. To avoid this problem the mask can be floated above the wafer in a process known as proximity printing, but at a cost of degraded resolution. Projection lithography systems capable of submicron resolution were introduced. To achieve increased resolution in either type of optical system, it is desirable to use shorter wavelengths of exposing radiation. Although mercury arc lamps have historically been the most widely used source, excimer lasers are dominant in current-generation and advanced lithography tools. Finally, methods to increase resolution through mask making were introduced, primarily the use of phase-shifted masks and optical proximity correction.

The chapter began by the observation that lithography plays a critical role in determining the performance of a technology. As such, lithography has long been the gating process in technology development. As a result, it is natural to wonder how far optical lithography can be pushed. Non optical techniques, several of which will be reviewed in Chapter 9, suffer from severe drawbacks compared to optical lithography. There is a considerable amount of truth to the suggestion that the limit of optical lithography is roughly three generations beyond the current state of the art, and has been for the last 20 years. It is expected that 193 nm sources, immersion lithography, and OPC will extend optical lithography to at least 45 nm, and perhaps to 30 nm. If large-scale optics can be created for the F_2 laser, optical lithography will probably be extended to 20 nm. Resist improvements and mask refinements such as phase shifting and OPC are decreasing the minimum feature size faster than the exposing wavelength.

PROBLEMS

1. What is role of lithographic step in integrated circuit manufacturing process? Explain in detail.
2. Write short note on following:
   (i) Optical Lithography
   (ii) Proximity optical lithography
3. Some arc lamps produce a significant amount of energy in the deep UV because of the high energy electrons in the plasma. Ozone creation is therefore, a significant concern. Calculate the plasma temperature required for the blackbody component of the radiation to be maximum at 250 nm.
4. A particular resist process is able to resolve features whose MTF is $\geq$ 0.4. Calculate the minimum feature size for an i-line aligner with an NA = 0.4 and S = 0.5.

REFERENCES

1. G. Stevens, "Microphotography," Wiley, New York, (1967).
2. M. Bowden, L. Thompson, and C. Wilson, "Introduction to Microlithography," American Chemical Society, Washington, DC, (1983).
3. D. Elliott, "Integrated Circuit Fabrication Technology," McGraw-Hill, New York, (1982).
4. W.M. Moreau, "Semiconductor Lithography, Principles, Practices, and Materials," Plenum, New York, (1988).
5. S. Nanogaki, T. Heno, and T. Ho, Microlithography Fundamentals in Semiconductor Devices and Fabrication Technology, Dekker, New York (1988).

References

6. P. Burggraaf, "Lithography's Leading Edge, Part 2: I-line and Beyond," *Semicond. Int.* 15(3):52 (1992).
7. The National Technology Roadmap for Semiconductors-1997, Semiconductor Industry Association, San Jose, CA, (1997).
8. M.V. Klein, "Optics," Wiley, New York, (1970).
9. M. Bowden and L. Thompson, "Introduction to Microlithography," American Chemical Society, Washington, DC, (1983).
10. K. Jain, "Excimer Laser Lithography," SPIE Optical Engineering Press. Bellingham, WA, (1990).
11. Malcolm Gower, "Excimer laser microfabrication and micromachining Excimer laser microfabrication and micromachining ," RIKEN Review No. 32 (January, 2001).
12. H. Craighead, J.C. White, R.E. Howard, L.D. Jackel, R.E. Behringer, J.E. Sweeney, and R.W. Epworth, "Contact Lithography at 157 nm with an F2 Excimer Laser," *J. Vacuum Sci. Technol.* B, 1:1186 (1983).
13. P. Concidine, "Effects of Coherence on Imaging Systems," *J. Opt. Soc. Am.* 56:1001 (1966).
14. J.H. Bruning, "Optical Lithography Below 100 nm," *Solid State Technol.* 41(11):59 (1998).
15. M.S. Hibbs, "Optical Lithography at 248 nm," *J. Electrochem. Soc.* 138:199 (1991).
16. A. Voschenkov and H. Herrman, "Submicron Resolution Deep UV Photolithography," *Electron. Lett.* 17:61 (1980).
17. A. Yoshikawa, S. Hirota, O. Ochi, A. Takeda, and Y. Mizushima, "Angstroms Resolution in Se-Ge Inorganic Resists," *Jpn J. Appl. Phys.* 20:L81 (1981).
18. H. Smith, "Fabrication Techniques for Surface-Acoustic-Wave and Thin-Film Optical Devices," *Proc. IEEE* 62:1361 (1974).
19. B. Lin, "Deep UV Lithography," J. Vacuum Sci. Technol. 12:1317 (1975).
20. G. Geikas and B.D. Ables, "Contact Printing - Associated Problems," Kodak Photoresist Seminar, 22 (1968).
21. B. Lin, R. Newman(ed.), "Fine Line Lithography," North-Holland, Amsterdam, 141 (1980).
22. J.E. Roussel, "Submicron Optical Lithography?" Semiconductor Microlithography, *Proc. SPIE* 275:9 (1981).
23. D.A. Markle, "A New Projection Printer," *Solid State Technol.*, 17:50 (1974).

24. M.C. King, "New Generation of Optical 1:1 Projection Aligners," Developments in Semiconductor Microlithography IV, *Proc. SPIE* 174:70 (1979).

25. R.T. Kerth, K. Jain, and M.R. Latta, "Excimer Laser Projection Lithography on a Full-Field Scanning Projection System," *IEEE Electron Dev. Lett.* EDL-7:299 (1986).

26. P. Burggraaf, "Wafer Steppers and Lens Options," *Semicond. Int.* 9:56 (1986).

27. K. Hennings and H. Schuetze, "Surface Complexation Parameters (SCP)," *Solid State Technol.* 31 (1966).

28. M.A. van den Brink, B.A. Katz, and S. Wittekoek, "A New 0.54 Aperture i-line Wafer Stepper with Field by Field Leveling Combined with Global Alignment," in Optical/Laser Microlithography IV, V Pol, ed., *Proc. SPIE* 1463:709 (1991).

29. R. Unger, C. Sparkes, P. DiSessa, and D. J. Elliott, "Design and Performance of a Production-Worthy Excimer-Laser-Based Stepper," in Optical/Laser Microlithography IV, V. Pol, ed., *Proc. SPIE* 1674:708 (1992).

30. B. Vleeming, B. Heskamp, H. Bakker, L. Verstappen, J. Finders, J. Stoeten, R. Boerret, and O. Roempp, "ArF Step-and-Scan System with 0.75 NA for the 0.10 µm node," Proc. SPIE 4346:634, Optical Microlithography XIV, Christopher J. Progler, ed., (2001).

31. Bernard Fay "Advanced Optical Lithography Development, from UV to EUV," *Microelectron Eng.* 61-62:11-24 (2002).

32. D. Gil, T. Brunner, C. Fonseca, and N. Seong, "Immersion Lithography: New Opportunities for Semiconductor Manufacturing," *J. Vacuum Sci. Technol.* B 22(6) (2004).

33. Nikon, "Immersion Lithography: System Design and Its Impact on Defectivity," (2005).

34. B. Smith, A. Bourov, Y. Fan, F. Cropanese, and P. Hammond, "Amphibian XIS: An Immersion Lithography Microstepper Platform," *Proc. SPIE* 5754 (2005).

35. M. Switkes, M. Rothschild, R. R. Kunz, S.-Y. Baek, and M. Yeung, "Immersion Lithography: Beyond the 65 nm Node with Optics," Microlithography World (May 2003); found at "Immersion Lithography," ICKnowledge.com (2003).

36. L. Geppert, "Chip Making's Wet New World," *IEEE Spectrum*, 41(5):21 (2004).

37. S. Owa, Y. Ishii, and K. Shiraishi, "Exposure Tool for Immersion Lithography," IEEE/SEMI Advanced Semiconductor Manufacturing Conference, (2005).

38. S. Peng, R. French, W. Qiu, R. Wheland, and M. Yang, "Second Generation Fluidsfor 193 nm Immersion Lithography," *Proc. SPIE* 5754 (2005).
39. J. Park, "The Interaction of Ultra-Pure Water and Photoresist in 193 nm Immersion Lithography," Microelectronic Engineering Conference, May (2004).
40. J. Taylor, Christopher; Shayib, Ramzy; Goh, Sumarlin; "Experimental Techniques for Detection of Components Extracted from Model 193 nm Immersion Lithography Photoresists," *Chem. Mater.* 17:4194 (2005).
41. M. Slezak, Z. Liu, and R. Hung, "Exploring the Needs and Tradeoffs for Immersion Resist Topcoating," *Solid State Technol.* (July 2004).
42. H. Sewell, D. McCafferty, L. Markoya, and M. Riggs, "Immersion Lithography, Next Step on the Roadmap," Brewer Science ARC Symposium, (2004).
43. B. Smith, A. Bourov, Y. Fan, F. Cropanese, and P. Hammond, "Air Bubble-Induced Light- Scattering Effect on Image Quality in 193 nm Immersion Lithography," *Appl. Opti.* 44:3904 (2005).
44. A. Wei, M. El-Morsi, G. Nellis, A. Abdo, and R. Engelstad, "Predicting Air Entrainment Due to Topography During the Filling and Scanning Process for Immersion Lithography," *J. Vacuum Scie. Technol.* B 22(6) (Nov/Dec 2004).
45. R. Unger and P. Disessa, "New i-line and Deep-UV Optical Wafer Steppers," in *Optical/Laser Microlithography IV*, V. Pol, ed., *Proc. SPIE* 1463:709 (1991).
46. R. Herschel, "Pellicle Protection of Integrated Circuit Masks," in *Semiconductor Microlithography VI*, *Proc. SPIE* 275:23 (1981).
47. P. Frasch and K. Saremski, "Feature Size Control in IC Manufacturing," *IBM J. Res. Dev.* 26:561 (1982).
48. B.J. Lin, "Phase-Shifting and Other Challenges in Optical Mask Technology," 10th Annu. Symp. Microlithography, *SPIE* 1496:54 (1990).
49. M.D. Levenson, N.S. Viswnathan, and R.A. Simpson, "Improving Resolution in Photolithography with a Phase Shifting Mask," *IEEE Trans. Electron Dev.* ED-26:1828 (1982).
50. G.E. Flores and B. Kirkpatrick, "Optical Lithography Stalls X-rays," *IEEE Spectrum* 28(10):24 (1991).
51. A.K. Pfau, W.G. Oldham, and A.R. Neureuther, "Exploration of Fabrication Techniques for Phase-Shifting Masks," in *Optical/Laser Microlithography IV*, V. Pol, ed., *Proc. SPIE* 1463:124 (1991).

52. A. Nitayama, T. Sato, K. Hashimoto, F. Shigemitsu, and M. Nakase, "New Phase-Shifting Mask with Self-Aligned Phase-Shifters for a Quarter-Micron Photolithography," *Tech. Dig. IEDM*, 1989, 3.3.1 (1989).

53. Y. Yanagishita, N. Ishiwata, Y. Tabata, K. Nakagawa, and K. Shigematsu, "Phase-Shifting Photolithography Applicable to Real IC Patterns," in *Optical/Laser Microlithography IV*, V. Pol, ed., *Proc. SPIE* 1463:124 (1991).

54. M.A. Listvan, M. Swanson, A. Wall, and S.A. Campbell, "Multiple LayerTechniques in Optical Lithography: Applications to Fine Line MOS Production," in *Optical Microlithography III: Technology for the Next Decade, Proc. SPIE* 470:85 (1983).

5

Etching

5.1 INTRODUCTION

Thin-film removal, also known as thin-film etch or just etch for short, is the process of selectively removing the unneeded (unprotected) material by a chemical (wet) or physical (dry) means. After a photoresist image has been fabricated on the surface of a wafer, the next process often involves transferring that image into a layer under the resist by etching. The chapter will begin with simple wet chemical etching processes where wafer is immersed in a solution that reacts with the exposed film to form soluble by-products. Ideally, the photoresist mask is highly resistant to attack by the etching solution. Although still used for noncritical processes, wet chemical etching is difficult to control, is prone to high defect levels due to solution particulate contamination, cannot be used for small features, and produces large volumes of chemical waste. The chapter will therefore go on to discuss dry or plasma etch processes. It is useful to begin the discussion of some important parameters related to the etch process by identifying the appropriate figures of merit.

5.2 ETCH PARAMETERS

The most important parameters that govern which of etching process, wet etching or dry etching, is appropriate when removing a specific thin film are etch rate, etch profile, selectivity, uniformity, and degree of anisotropy.

Etch Rate

Etch rate is a measure of the thickness removed per unit of time. It is usually a strong function of solution concentration and etching temperature. High etch rate is generally favorable due to a higher throughput which is generally desirable in a manufacturing environment. Too high etch rate may render a process difficult to control. Desired etch rates commonly are hundreds or tens of nanometer per minute (nm/min). When a batch of wafers is etched simultaneously, etch rate can be less than the rate required for a single-wafer etch process. Several related figures are equally important. Variation percentage of each rate through a wafer and wafer to wafer is measured in terms.

Etch Profile

Etch profile refers to the fraction of sidewall of the etched feature removed during an etching process. There are two basic etch profiles, as illustrated in figure 5.1. In an isotropic etch profile, all directions are etched at the same rate, leading to an undercut of the etched material under the mask as in figure 5.1(a). This results in the reduction of the actual width of a line such as a polysilicon or a metal wire. The other etch profile is called an anisotropic etch profile. In this profile, the etch rate is in only one direction perpendicular to the wafer surface, as in figure 5.1(b).

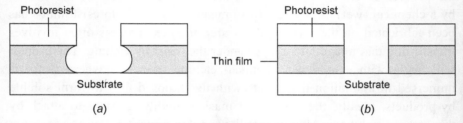

Fig. 5.1 (a) Isotropic (b) Anisotropic

Selectivity

Selectivity means how much faster one material is etched than another under the same condition. Selectivity (S) is defined as etch rate ratio of one material to another and is given by

$$S = \frac{R_1}{R_2} \qquad \ldots (5.1)$$

where R_1 is etch rate of the material intended to be removed and R_2 is the etch rate of the material not intended to be removed. A particular process may be quoted as having a selectivity of 20 to 1 for polysilicon over oxide means that polysilicon etches 20 times faster than oxide.

Uniformity

Uniformity is a measure of the capability of the etching process to etch evenly across the entire wafer surface. For IC production lines, high uniform each rates are important. Each rate uniformity is given by

$$\text{Each rate uniformity (\%)} = \frac{(\text{Maximun etch rate} - \text{Minimum etch rate})}{(\text{Maximum etch rate} + \text{Minimum etch rate})} \times 100$$

Degree of Anisotropy

Degree of anisotropy A_f is a measure of how rapidly an etchant removes material in different directions and can be given by

$$A_f = 1 - \frac{R_l}{R_v} \qquad \ldots (5.2)$$

where R_l is the lateral etch rate whereas R_v is the vertical etch rate. For isotropic etch, R_l is equal to R_v and therefore $A_f = 0$; for anisotropic etch, R_l is equal to 0 and therefore $A_f = 1$.

5.3 WET ETCHING PROCESS

The wet etching process is the earliest used etching process. It uses the chemical reaction between thin film and solvent to remove the thin film unprotected by photoresist. The wet etching process is commonly used to etch silicon dioxide, single-crystal silicon, silicon nitride, and metal. The wet etching process has high throughput compared to the dry etching process and is usually an isotropic process even though it can also be anisotropic. Consequently, in modern deep submicron processes, it is not suitable for defining the line features. However, due to high selectivity, it still plays an important role in the cleaning of the wafer surface and thin-film removal, such as silicon dioxide cleaning, residue removal, and the stripping of surface layers, such as the blanket thin film. So in wafer production process lapping and polishing process utilizes wet etching to produce damage-free optically flat surface. Wafers are scrubbed and chemically cleaned to remove contamination that results from handling and storing before loading to the chambers for thermal oxidation or epitaxial growth. Wet etching is used to outline patterns and windows in insulating materials for many discrete devices and ICs of relatively large dimensions (> 3 µm).

In IC processing, most chemical etchings proceed through dissolution of a materials in a solvent or by transforming the material into a soluble compound that successively dissolves in the etching medium. Wet chemical etching contains mainly three steps:

1. Transportation of reactants towards the reacting surface (e.g. by diffusion)
2. Chemical reactions on and at the surface
3. Transportation of the products from the surface (e.g. by diffusion)

Since all three steps must occur, the slowest one, called the rate-limiting step, determines the etch rate. Since it is generally desirable to have a large, uniform, well-controlled etch rate, the wet etch solution is often agitated in some manner to assist in the movement of etchant to the surface and the removal of the etch product. Some wet etch processes use a continuous acid spray to ensure a fresh supply of etchant, but this comes at the cost of the production of significant amounts of chemical waste. Wet etching can also have serious drawbacks such as a lack of anisotropy, poor process control, and excessive particle contamination.

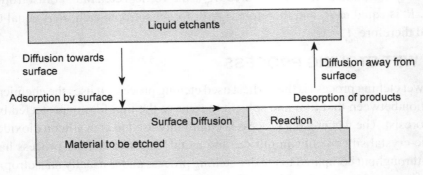

Fig. 5.2 Wet Etching Mechanism

For most wet etch processes, the film to be etched is not directly soluble in the etchant solution. It is usually necessary to use a chemical reaction to change the material to be etched from a solid to a liquid or a gas. If the etching process produces a gas, this gas can form bubbles that can prevent the movement of fresh etchant to the surface. This is an extremely serious problem, since the occurrence of the bubbles cannot be predicted. The problem is most pronounced near pattern edges. In addition to assisting the movement of fresh etchant chemicals to the surface of the wafer agitation in the wet chemical bath will reduce the ability of the bubbles to adhere to the wafer. Even in the absence of bubbles, however, small geometry features may etch more slowly, due to the difficulty in removing all of the etch products. This phenomenon has been shown to be related to microscopic bubbles of trapped gas. Another common problem for wet etch processes is undetected resist scumming. This occurs when some of the exposed photoresist is not removed in the develop process. Common causes are incorrect or incomplete exposures and insufficient developing of the pattern. Due to the high selectivity of wet etch processes, even a very thin layer of resist residue is sufficient to completely block the wet etch process.

In the 1990s, wet etching enjoyed something of resurgence. Automated wet etch benches were developed that allow the operator to precisely control the etch time, bath temperature, degree of agitation, bath composition, and the degree of misting in spray etches. Increased use of filtration, even in hot, very aggressive, compounds, has helped control particle deposition concerns. Even with these improvements, however, wet etching is still not regarded as practical for most features smaller than 2 μm.

5.4 SILICON ETCHING

For silicon, mostly used etchants are the mixtures of nitric acid (HNO_3) with hydrofluoric acid (HF) in water or acetic acid (CH_3COOH). The reaction is initiated by promoting silicon from its initial oxidation state to a higher oxidation state

$$Si + 2h^+ \longrightarrow Si^{2+}$$

The holes (h^+) are generated through the following autocatalytic process:

$$HNO_3 + HNO_2 \longrightarrow 2NO_2^- + 2h^+ + H_2O$$
$$2NO_2^- + 2H^+ \longrightarrow 2HNO_2$$

Si^{2+} combines with OH^- (produced by the dissociation of H_2O) to form $Si(OH)_2$ which later releases H_2 to form SiO_2:

$$Si(OH)_2 \longrightarrow SiO_2 + H_2$$

SiO_2 then dissolves in HF;

$$SiO_2 + 6HF \longrightarrow H_2SiF_6 + H_2O$$

The overall reaction can be written

$$Si\,(s) + HNO_3\,(l) + 6HF\,(l) \longrightarrow H_2SiF_6\,(l) + HNO_2\,(l) + H_2O\,(l) + H_2\,(g)$$

Although water can be used as a diluent, acetic acid is preferred as the dissociation of nitric acid can be retarded in order to yield a higher concentration of the undissociated species at very high HF and low HNO_3 concentrations, the etching rate is controlled by HNO_3, because there is an excess amount of HF to dissolve any SiO_2 formed at low HF and high HNO_3 concentrations, the etch rate is controlled by the ability of HF to remove SiO_2 as it is being formed. The later etching mechanism is isotropic, that is not sensitive to crystallographic orientation. Figure 5.3 shows the etch rate of silicon in HF and HNO_3. Notice that the three axes are not independent. To find the etch rate, draw lines from the percentage of HNO_3 and HF. They should intersect at a point on the line corresponding to the remaining percentage of diluent at low HNO_3 concentrations, the etch rate is controlled by the oxidant concentration. At low HF concentrations, the etch rate is controlled by the HF concentration. The maximum etch rate for this solution is 470 μm/min. A hole can be etched completely through a typical wafer at this rate in about 90 sec.

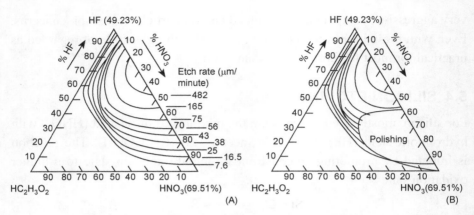

Fig. 5.3 The etch rate of Si in HF and HNO$_3$

Some etchant dissolves a given crystal plane of single-crystal silicon with much faster than another plane which results in orientation-dependent etching. For a silicon lattice, the (111) plane has more available bonds per unit area than the (110) and (100) planes so expected to be slower for the (111) plane A commonly dependent etch for silicon consists of a mixture of KOH in water and isopropyl alcohol. The {100}-and {110}-crystal planes are being etched, the stable {111} planes act as an etch stop (111)-orientated Si-wafers are almost not attacked by the etch (100)-orientated wafers form square-based pyramids with {111} surfaces. These pyramids are realized on C-Si solar

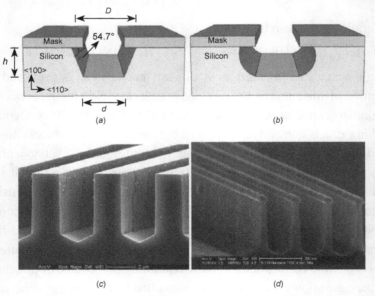

Fig. 5.4 Schematic OD etch profile in a) an anisotropic and (b) an isotropic of a (100) oriented silicon surface (c-d) SEM image of KOH based wet etching for (110) - oriented Si surface

cells for the purpose of reflection minimization (110)-orientated wafers form perpendicular trenches with {111} side-walls, used as e.g. micro channels in micro mechanics and micro fluidics Etching a (100) silicon wafer would result in a pyramid shaped etch pit as shown in Figure 5.4 (a). The etched wall will be flat and angled. The angle to the surface of the wafer is 54.7°. Figure 5.4 (c and d) depicts scanning electron micrographs of (110)-oriented two-dimensional silicon walls with micro and nano scale dimensions generated based on KOH based wet etching The relationship between mask dimensions, etch depth and the floor width is given in equation.

$$d = D - \frac{2h}{\tan(54.7°)} \quad \ldots (5.3)$$

5.5 SILICON DIOXIDE ETCHING

One of the most common etching processes is the wet etching of SiO_2 in dilute solutions of hydrofluoric acid (HF). Common etchants are 6:1, 10:1, and 50:1, meaning 6, 10, or 50 parts (by volume) of water to one part HF. A 6:1 HF solution will etch thermal silicon dioxide at about 1200 Å/min. Deposited oxides tend to etch much faster. The ratio of the deposited film etch rate in HF to that of thermal oxides is often taken as a measure of its density. Doped oxides such as phosphosilicate glass and borophosphosilicate glass etch faster yet, as the etch rate increases with impurity concentration. Solutions of HF are extremely selective of oxide over silicon some etching of silicon does occur, since the water will slowly oxidize the surface of the silicon and HF will etch this oxide Selectivity's are commonly better than 100:1. Wet etching of oxide in HF solutions is, however, completely isotropic. The exact reaction pathway is complex and depends on the ionic strength, the solution pH, and the etchant solution. The overall reaction for etching SiO_2 is

$$SiO_2 (s) + 6HF (l) \longrightarrow H_2 (g) + SiF_6 (l) + 2H_2O (l)$$

Since the reaction consumes HF, the reaction rate will decrease with time. To avoid this it is common to use HF with a buffering agent (BHF) such as ammonium fluoride (NH_4F), which maintains a constant concentration of HF through the dissolution reaction:

$$NH_4F (s) \longleftrightarrow NH_3 (g) + HF (l)$$

Where NH_3 (ammonia) is a gas. Buffering also controls the pH of the etchant which minimizes photoresist attack.

Silicon Nitride

Silicon nitride is etched very slowly by HF solutions at room temperature. A 20:1 BHF solution at room temperature, for example, etches thermal oxide at about 300 Å/min, but the etch rate for Si_3N_4 is less than 10 Å/min. More practical etch rates of Si_3N_4 can be obtained in H_3PO_4 at 140 to 200°C. A 3:10 mixture of 49% HF (in H_2O) and 70% HNO_3 at 70°C can also be used, but is much less common. Typical selectivities in the phosphoric etch are 10:1 for nitride over oxide and 3:1 for nitride over silicon. If the nitride layer is exposed to a high temperature oxidizing ambient, therefore, a dip in BHF is often done before the nitride wet etch to strip any surface oxide that may have grown on top of the nitride. Better patterning can be achieved by depositing a thin oxide layer on top of the nitride film before resist coating. The resist pattern is transferred to the oxide layer, which then acts as a mask for subsequent nitride etching.

5.6 ALUMINUM ETCHING

Wet etching has also been widely used to pattern metal lines. Since the metal layers used in ICs are often polycrystalline, the lines produced by wet etching sometimes have ragged edges. A common aluminum etchant is (by volume) 20% acetic acid, 77% phosphoric acid and 3% nitric acid. Most of the metal interconnects in silicon technologies is not elemental but rather dilute alloys. In many cases, these impurities are much less volatile in the bath than the base material. In particular, silicon and copper additions to aluminum are often difficult to completely remove in standard aluminum wet etch solutions.

Etching of insulating and metal films is usually performed with the same chemicals that dissolve these materials in bulk form and involves their conversion into soluble salts and complexes (Table 5.1). Generally speaking, film materials will etch more rapidly than their bulk counterparts. Moreover, the etching rates are higher for films that possess poor microstructures or built-in stress, are non-stoichiometric, or have been irradiate.

Table 5.1 Etch Rate and Composition

Material	Etchant		Etch Rate
SiO_2	28 ml HF 170 ml H_2O 113 g NH_4F	Buffered HF (BHF)	100 nm/min
	15 ml HF 10 ml HNO_3 300 ml H_2O	P – Etch	12 nm/min

Material	Etchant	Etch Rate
Si_3N_4	Buffered HF H_3PO_4	0.5 nm/min 10 nm/min
Al	1 ml HNO_3 4 ml CH_3COOH 4 ml H_2PO_4 1 ml H_2O	35 nm/min
Au	4 g KI 1 g I_2 40 ml H_2O	1 μm/min
Mo	5 mg H_3PO_4 2 ml HNO_3 4 ml CH_3COOH 150 ml H_2O	0.5 nm/min
Pt	1 ml HNO_3 7 ml HCl 8 ml H_2O	50 nm/min
W	34 g KH_2PO_4 13.4 $K_3Fe(CN)_6$ H_2O to make 1 liter	160 nm/min

5.7 DRY ETCHING PROCESS

The dry etching (also called as plasma etching) process has gradually replaced the wet etching process for all patterned etching processes since the feature size reached 3 μm in the late 1980s. Nowadays, because it has an excellent anisotropic profile and can generate very reactive chemical species, the dry etching process has become the primary etch approach in semiconductor fabricating. It may remove the material through only chemical reactions using chemical reactive gases or plasma, by purely physical methods such as sputtering and ion beam-induced etching, or with a combination of both chemical reaction and physical bombardment. The dry etching process is commonly used to etch dielectric, single-crystal silicon, poly silicon, and metal, as well as to strip photoresist.

5.8 PLASMA ETCHING PROCESS

Plasma is an ionized gas composed of ions, electrons, and neutral atoms or molecules with an equal amount of positive and negative charge. Although plasma is neutral in a macroscopic sense, it behaves quite differently from a molecular gas, because it consists of charged particles that can be influenced by applied electric and magnetic fields. To achieve the etching action, plasma provides energetic positive ions that are accelerated toward the wafer surface by a high electric field. These ions physically bombard the unprotected wafer surface material, causing material to be ejected off the wafer surface. A plasma is produced when an electric field is applied across two electrodes between which a gas is confined at low pressure, causing the gas to break down and become ionized simple DC (direct current) power can be used to generate plasma, but insulating materials require AC (alternate current) power to reduce charging. In plasma etching, an RF (radio frequency) field is usually used to generate the gas discharge. One reason for doing so is that the electrodes do not have to be made of a conducting material. The other reason is that electrons can pick up sufficient energy during field oscillation to cause more ionization by electron – neutral atom collisions. As a result, the plasma can be generated at pressures lower than 10^{-3} torr. A conceptual view of the plasma etching system is shown in figure 5.5.

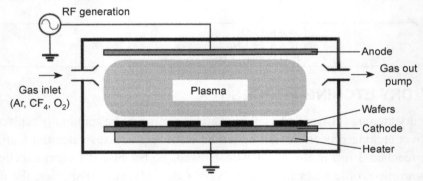

Fig. 5.5 Schematic view of plasma etching system

The free electrons released by photo-ionization or field emission from a negatively biased electrode create the plasma. The free electrons gain kinetic energy from the applied electric field, and in the course of their travel through the gas, they collide with gas molecules and lose energy. These inelastic collisions serve to further ionize or excite neutral species in the plasma via the following reaction examples:

$$e^- + AB \longrightarrow A^- + B^+ + e^- \text{ (Dissociative attachment)}$$
$$e^- + AB \longrightarrow A + B + e^- \text{ (Dissociation)}$$
$$e^- + A \longrightarrow A^+ + 2e^- \text{ (Ionization)}$$

Some of these collisions cause the gas molecules to be ionized and create more electrons to sustain the plasma. Therefore, when the applied voltage is larger than the breakdown potential, the plasma is formed throughout the reaction chamber. Some of these inelastic collisions can also raise neutrals and ions to excited electronic states that later decay by photoemission, thereby causing the characteristic plasma glow.

5.8 Plasma Etching Process

The interaction of plasmas with surfaces is often divided into two components physical and chemical. A physical interaction refers to the surface bombardment of energetic ions accelerated across the plasma sheath. Here the loss of kinetic energy by the impinging ions causes ejection of particles from the sample surface. Conversely, chemical reactions are standard electronic bonding processes that result in the formation or dissociation of chemical species on the surface.

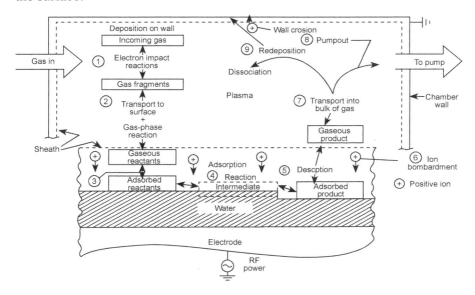

Fig. 5.6 Schematic view of microscopic processes that takes place during plasma etching of Silicon wafer

As exhibited in figure 5.6, the plasma-assisted etching process proceeds in several steps. It commences with the generation of the etchant species in the plasma, accompanied by diffusion of the reactant through the dark sheath to the specimen surface. After the reactant adsorbs on the surface, chemical reactions and/or physical sputtering occur to form volatile compounds and/or atoms that are subsequently desorbed from the sample surface, diffuse into the bulk gas, and pumped out by the vacuum system.

There are two main types of species involved in plasma etching: the reactive neutral chemical species and the ions. The reactive neutral chemical species are mainly responsible for the chemical component in the plasma etching process while the ions are for the physical component. According to whether these two components work together or independently, the plasma etching process can be differentiated into three processes: the plasma chemical etching process, sputter etching process, and reactive ion etch (RIE) process. All of these are the most important dry etching processes used in the IC industry nowadays.

5.8.1 Plasma Chemical Etching Process

The plasma etching process is also referred to as the plasma chemical etching process since in such a system, the material to be etched is commonly removed by a chemical reaction between it and the reactive chemical species. This process involves a chemical reaction between etchant gases to attack the silicon surface. The chemical dry etching process is usually isotropic and exhibits high selectively. Anisotropic dry etching has the ability to etch with finer resolution and higher aspect ratio than isotropic etching. Due to the directional nature of dry etching, undercutting can be avoided. The reactive chemical species are free radicals, which is neutral species having incomplete bonding. Due to their incomplete bonding structures, free radicals are very highly reactive chemical species. The species used in the plasma etching process need to have the property that it will generate a volatile by-product when reacting with the material to be etched. Thereby, the by-product can be easily removed and the surface is made to expose for more material to be etched. Due to the chemical reaction involved in this process, it has high selectivity but poor direction, i.e., an isotropic profile. Figure 5.7 shows a rendition of the reaction that takes place in chemical dry etching. Some of the ions that are used in chemical dry etching is carbon tetra fluoride (CF_4), sulfur hexafluoride (SF_6), nitrogen tri fluoride (NF_3), chlorine gas (Cl_2), or fluorine (F_2). There is a wide variety of dry etch processes with differing amounts of physical and chemical attack. Overlaid on this is the variety of etch chemistries that is used in each type of etch system. Table 5.2 lists some of the most common.

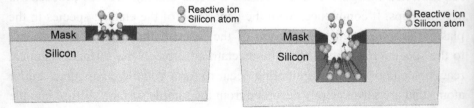

Fig. 5.7 Process of a reactive ion interacting with the silicon surface. Interaction bonding then chemically removal

Table 5.2 Typical etch chemistries

Si	CH_4/O_2, CF_2Cl_2, CF_3Cl, $SF_6/O_2/Cl_2$, $Cl_2/H_2/C_2F_6/CCl_4$, C_2ClF_5/O_2, Br_2, SiF_4/O_2, NF_3, ClF_3, CCl_4, CCl_3F_5, C_2ClF_5/SF_6, C_2F_6/CF_3Cl, CF_3Cl/Br_2
SiO_2	CF_4/H_2, C_2F_6, C_3F_8, CHF_3/O_2

5.8 Plasma Etching Process

Si_3N_4	$CF_4/O_2/H_2$, C_2F_6, C_3F_8, CHF_3
Organics	O_2, CF_4/O_2, SF_6/O_2
Al	BCl_3, BCl_3/Cl_2, $CCl_4/Cl_2/BCl_3$, $SiCl_4/Cl_2$
Silicides	CF_4/O_2, NF_3, SF_6/Cl_2, CF_4/Cl_2
Refractories	CF_4/O_2, NF_3/H_2, SF_6/O_2
GaAs	BCl_3/Ar, $Cl_2/O_2/H_2$, $CCl_2F_2/O_2/Ar/He$, H_2, CH_4/H_2, $CClH_3/H_2$
InP	CH_4/H_2, C_2H_6/H_2, Cl_2/Ar
Au	$C_2Cl_2F_4$, Cl_2, $CClF_3$

5.8.2 Sputter Etching Process

Another plasma etching process makes use of the ions; that ions are responsible for the etching operation. In this etching process, ions accelerated by a high electric field bombard the atoms on the wafer surface so as to physically dislodge them. This bombardment results in more physical components of etching, that is, sputtered surface materials. When the high energy particles knock out the atoms from the substrate surface, the material evaporates after leaving the substrate. There is no chemical reaction taking place and therefore only the material that is unmasked will be removed. The physical reaction taking place is illustrated in figure 5.8. Because it is fundamentally a physical reaction, it has poor selectivity but high direction, i.e., an anisotropic profile.

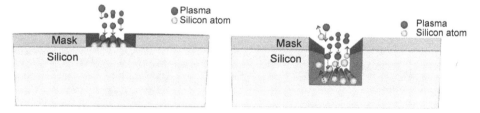

Fig. 5.8 Sputtering Etch System

Figure 5.9 is the schematic of a sputtering-etching system utilizing relatively high energy (> 500 eV) noble gas ions such as argon. The wafer to be etched (called target) is placed on a powered electrode, and argon ions are accelerated by the applied electric field to bombard the target surface. Through the transfer of momentum, atoms near the surface are sputtered off the surface. The typical operating pressure for sputter etching is 0.01 to 0.1 torr. The direction of the electric field is normal to the target surface and under the operating pressure, argon ions arrive predominantly normal to the surface. Consequently, there is essentially no sputtering of the sidewalls and a high degree of anisotropy can be attained. However, a major drawback for sputter etching is its poor selectivity, for the ion bombardment process etches everything on the surface, albeit the difference in sputtering rates for different materials.

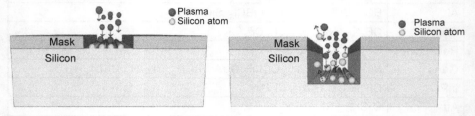

Fig. 5.9 Schematic of a sputtering - etching system

5.8.3 Reactive Ion Etching (RIE) Process

An alternative plasma etching method is reactive ion etching (RIE), which employs apparatus similar to that for sputter etching shown in figure 5.10. The primary difference here is that the noble gas plasma is replaced by molecular gas plasma similar to that in plasma etching. The approach is that both free radicals and ions work together in a synergistic manner to etch the material. In other words, the etching process involves both ion sputtering and radicals reacting with the wafer surface. The result not only has a high degree of selectivity but also achieves a very anisotropic etch profile. In this etching process, the actual etch profile is between isotropic and anisotropic and can be controlled by adjusting the plasma conditions and gas composition. The typical RIE gasses for Si are CF_4, SF_6 and $BCl_2 + Cl_2$.

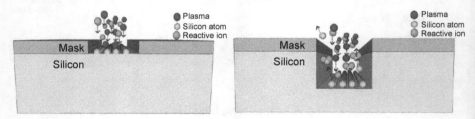

Fig. 5.10 RIE process which involves both physical and chemical reaction to etch of silicon

RIE uses chemically reactive plasma generated under a low pressure (10-100 m.torr) to consume the materials deposited on wafers, along with ionic bombardment which can open areas for chemical reactions. In the RIE systems, the chuck holding wafer is grounded and another electrode is connected to the radio frequency power of frequency 13.56 MHz since electrons are more mobile compared to positive ions due to their lighter weight, they travel longer and collide more frequently with the electrodes and chamber walls and consequently be removed from the plasma. This process leaves the plasma

5.9 Inductive coupled Plasma Etching (ICP)

positively charged. Nevertheless, plasma tends to remain neutral charged thus, a DC electrical field is formed. The region of surfaces in the chamber and electrodes as shown in figure 5.11 is called "dark sheath". Besides the chemical reaction between the plasma and the target material, the positive ions can be accelerated across the dark sheath by electrical field and strike on the target material. This process is the physical bombardment which can also assist the etching process. In addition, in contrast to the wet chemical RIE has a higher probability to move the etchants in the direction given by the electric field and produces more anisotropic etch profiles.

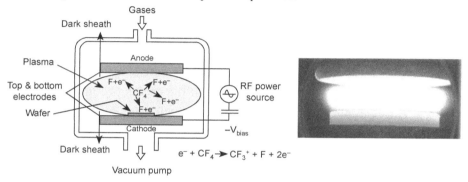

Fig. 5.11 RIE system with dark sheath area

5.9 INDUCTIVE COUPLED PLASMA ETCHING (ICP)

The ICP employs radio frequency energy generated by the electromagnetic induction coil to create plasma of ionized atoms and radicals capable of etching various semiconductor materials. As to the geometry and operation of ICP, the induction coil of the ICP is wrapped two or three times around the torch and has water flowing through it for cooling purposes. In order for the RF power to travel along the surface of the coil with minimum resistance, all ICPs have a capacitor bank which is continuously tuned to match the plasma's inductance. The RF power supply maintains the plasma and the tesla coil is used to ignite the plasma through the generation of electrons and ions, which couple with magnetic field. Very high plasma densities can be achieved by ICP and the etch profiles tend to be more isotropic than RIE.

Currently, the combination of typical RIE and inductively coupled plasma RIE is possible. In this system, the ICP is employed as a high density source of ions which increases the etch rate, whereas a separate RF bias is applied to the substrate to create directional electric fields near the substrate to achieve more anisotropic profiles, however, it can also be run in RIE mode for certain low etch rate applications and control over selectivity and damages.

5.10 ADVANTAGES AND DISADVANTAGES OF DRY ETCHING (PLASMA ETCHING) AND WET ETCHING

The advantages of wet etching processes are the simple equipment, high etching rate, and high selectivity. However, there are many disadvantages. Wet etching is generally isotropic, which results in the etchant chemicals removing substrate material under the masking material. Wet etching also requires large amounts of etchant chemicals because the substrate material has to be covered with the etchant chemical. Furthermore, the etchant chemicals have to be consistently replaced in order to keep the same initial etching rate. As a result, the chemical and disposal costs associated with wet etching are extremely high. Some advantages of dry etching are its capability of automation and reduced material consumption. Dry etching (e.g. plasma etching) costs less to dispose of the products compared to wet etching. An example of purely chemical dry etching is plasma etching. A disadvantage of purely chemical etching techniques, specifically plasma etching processes, is that they do not have high anisotropy because reacting species can react in any direction and can enter from beneath the masking material. Anisotropy is when the etching exclusively occurs in one direction. This property is useful when it is necessary to remove material only in the vertical direction since the material covered by the masking material would not be removed. In cases where high anisotropy is vital, dry etching techniques that use only physical removal or a combination of both physical removal and chemical reactions are used.

5.11 EXAMPLES OF ETCHING REACTIONS

In both plasma and reactive ion etching, ions from the plasma are attracted to the sample surface. However, the pure sputtering process is quite slow. The etching rate can be enhanced substantially by ion-assisted chemical reaction. Figure 5.12 depicts the etch rate as a function of flow rate of XeF_2 molecules with and without 1 keV Ne^+ bombardment. The lateral etch rate depends only on the ability of XeF_2 molecules to etch silicon in the absence of energetic ions impacting the surface, whereas the vertical etch rate is a synergistic effect due to both Ne^+ bombardment and XeF_2 molecules. The degree of anisotropy can generally be enhanced by increasing the energy of the ions.

5.11 Examples of Etching Reactions

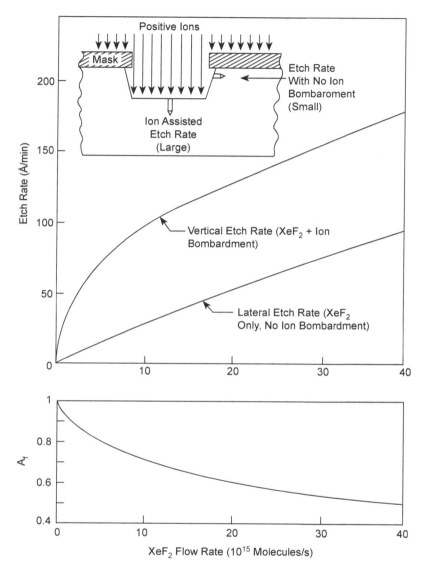

Fig. 5.12 Silicon etching rate vs XeF$_2$ flow rate with and with 1-KeV Ne$^+$ bombardment

When a gas is mixed with one or more additive gases, both the etch rate and selectivity can be altered. As illustrated in figure 5.13, the etching rate of SiO$_2$ is approximately constant for addition of up to 40% H$_2$, while the etch rate for Si drops monotonically and is almost zero at 40% H$_2$. Also shown is the selectivity, that is, the ratio of the etch rate SiO$_2$ to that of Si, Selectivity exceeding 45:1 can be achieved with CF$_4$-H$_2$ reactive ion etching. This process is thus useful when etching a SiO$_2$ layer that covers a poly-Si gate.

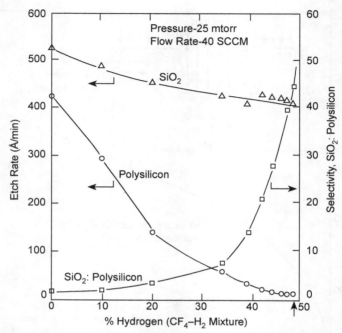

Fig. 5.13 Etching rate of Si and SiO_2 and the corresponding selectivity as a function Of mixture of H_2 and CF_4.

The opposite effect can be observed by varying the gas composition of sulfur hexafluoride (SF_6) and chlorine, as exhibited in figure 5.14. The etching rate of silicon can be adjusted to be 10 to 80 times faster than that of SiO_2. Examples of some common etchants exhibiting selectivity effects are exhibited in Table 5.3.

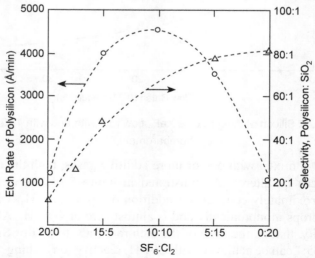

Fig. 5.14 Dependence of etching rate of poly silicon and selectivity of SiO_2 as a function of gas composition in $SF_6 - Cl_2$

Table 5.3 Etch rate and selectivity for dry etching

Material (M)	Gas	Etch Rate (Å/min)	Selectivity		
			M/Resist	M/Si	M/SiO$_2$
Si	SF$_6$ + Cl$_2$	1000–4500	5	–	80
SiO$_2$	CF$_4$ + H$_2$	400–500	5	40	–
Al, Al-Si, Al-Cu	BCl$_3$ + Cl$_2$	500	5	5	25

5.12 LIFTOFF

Most GaAs technologies were developed around liftoff rather than etching. The process is still popular for patterning difficult to etch materials. The sequence for liftoff is shown in figure 5.15. A thick layer of resist is spun and patterned. Next, a thin layer of metal is deposited using evaporation as discussed in metallization chapter (Chapter 8). One characteristic of evaporation is its difficulty in covering high aspect ratio features. If a reentrant profile is obtained in the resist, a break in the metal is virtually assured. Next, the wafer is immersed in a solution capable of dissolving the photoresist. The metal lines that were deposited directly on the semiconductor remain, while the metal deposited on the resist lifts off of the wafer as the resist dissolves. Etch damage to the substrate is avoided and the lines patterned with infinite selectivity with no undercut. Since the process in its simplest form requires only a wet bench and perhaps ultrasonic agitation, it is widely used in research laboratories.

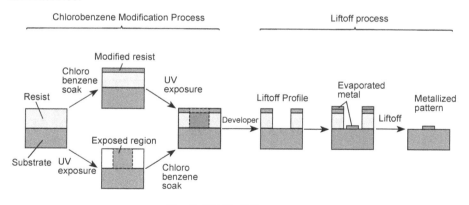

Fig. 5.15 Liftoff Sequence

Methods that have been used to form a reentrant resist profile generally harden the surface of the resist. This can be done to some extent by promoting cross-linking by deep UV exposure, in a suitable plasma environment, or by ion implantation. Another solution is the use of multiple layers of resist such as a DQN resist on PMMA. After PMMA coating, the upper layer of resist is spun, baked, exposed, and developed as normal with a UV source. The patterned upper layer can then be used as a mask for a deep UV exposure of the PMMA, which is then overdeveloped using a solution that does not attack the upper resist. The result is a pronounced ledge that is very difficult to cover. Due to the complexity of multilayer resist processing, the most popular method of producing reentrant profiles is soaking a single-layer DQN resist after soft bake in chlorobenzene or similar compounds. Typical soak times are 5 to 15 min. The soak process reduces the dissolution rate of the upper surface of the resist. After developing the pattern, therefore, a ledge appears (Figure 5.16). The thickness of the ledge depends on the soak time, the temperature of the chlorobenzene bath, and the resist prebake cycle.

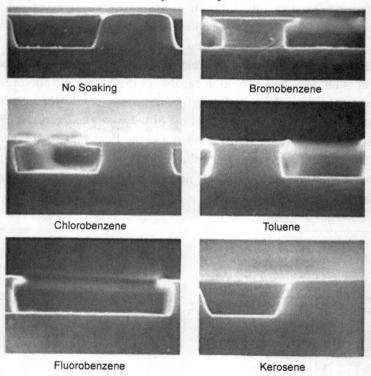

Fig. 5.16 Liftoff profiles after various resist treatment

Liftoff processes have several shortcomings. The first is that the surface topology must be very smooth, since the metal deposition step is designed to have poor step coverage. Therefore, either the technology must be limited to one layer of metallization, or each layer must be plenaries before liftoff patterning .This effectively prevents the use of sputtering. The other serious problem is that the metal lifted off remains solid and floats in the bath. Pieces of it are very likely to redeposit on the surface of the wafer. Unless the patterns are very simple, liftoff has serious yield impacts.

5.13 SUMMARY

Etching is a process to transfer patterns in IC fabrication. Thin film removal, also known as thin-film etch as just etch for short, is the process of selectively removing the unneeded (unprotected) material by a chemical (wet) or physical (dry) means selectivity & anisotrophy are two most importants issues. Wet chemical etching was firstly used suspensively in semiconductors processing. The wet chemical etching process for insulators, silicon and metal intercommunication are discussed in this chapter. Whereas to achieve high fidelity pattern transfer dry etching method is more preferable. Dry etching is used almost exclusively today because of the control, flexibility reproducibility and anisotropic behaviour. This etching is synonym with plasma assisted etching. Various dry etching system and plasma fundamentals are discussed in this chapter. The challenges for future etching technologies are low aspect ratio depended etching, better dimensional control, high etch selectivity and low plasma induced damage. To meet these requirements high density, low pressure plasma reactors are needed.

PROBLEMS

1. What are the major distinctions between reactive ion etching and parallel plate plasma etching?
2. Compare at the advantages and limitation of dry and wet etching techniques.
3. Explain various types of etching process used in integrated circuit manufacturing process.
4. Explain why endpoint detection by monitoring of reactant species requires a lading effect.
5. A cylindrical parallel-plate RF etch chamber is constructed with 14" diameter electrodes. The chamber diameter is 16" and the height of the chamber is 6". One of the electrodes is attached to the ground. The DC voltage between the electrodes is measured to be 25 V when a plasma is established. Assuming that the plasma is in contact with the

chamber walls, calculate the potential difference between the plasma and each electrode. Explain the significance of the result.

6. Describe the differences between high pressure plasmas and reactive ion-etch systems. Explain when each is the preferred process.

7. Describe the difference between chemically assisted ion beam etching (CAIBE) and ion-assisted chemical etching.

8. A solution consisting of four parts 5% HNO_3, four parts 49% HF, and two parts $C_2H_3O_2$ is used to etch silicon. If the solution is held at room temperature, what etch rate would you expect? If the compound is to remain at two parts $C_2H_3O_2$, what mixture of these same chemicals would be suitable to etch silicon at a rate of approximately 10 µm/min?

REFERENCES

1. W.A. Kern and C.A. Deckert, "Chemical Etching," Thin Film Processing, Academic press, New York, (1978).

2. K. McAndrews and P.C. Subanek, "Nonuniform Wet Etching of Silicon Dioxide," *J. Electrochem. Soc.* 138: 863 (1991).

3. P. Burggraaf, "Wet Etching: Alive, Well, and Futuristic," *Semicond. Int.* 13(9):58 (1990).

4. W. Kern, H.G. Hughes and M.J. Rand, "Chemical Etching of Dielectrics," Etching for Pattern Definition, Electrochemical Society, Pennington, NJ, (1976).

5. S.M. Hu and D.R. Kerf, "Observation of Etching of n-Type Silicon in Aqueous HF Solutions," *J. Electrochem. Soc.* 114:414 (1967).

6. J.S. Judge, H.G. Hughes and M.J. Rand, "Etching for Pattern Definition," Electrochemical Society, Princeton, NJ, (1976).

7. L.M. Loewenstein and C.M. Tipton, "Chemical Etching of Thermally Oxidized Silicon Nitride: Comparison of Wet and Dry Etching Methods," J. Electrochem. Soc. 138:1389 (1991).

8. J.T. Milek, "Silicon Nitride for Microelectronic Applications, Part 1—Preparation and Properties," IFI/Plenum, New York, (1971).

9. B. Schwartz and H. Robbins, "Chemical Etching of Silicon: Etching Technology," *J. Electrochem. Soc.* 123:1903 (1976).

10. Kern and Deckert "A comprehensive listing of etching solutions for groups III-V", Eds Academic Press, Enlands (1978).

11. R.E. Williams, "Gallium Arsenide Processing Techniques," Artech, Dedham, MA, (1984).

12. S. Adache and K. Oe, "Chemical Etching Characteristics of (001) GaAs," *J. Electrochem. Soc.* 130:2427 (1983).

13. Y. Tarui, Y. Komiya, and Y. Harada, "Preferential Etching and Etched Profiles of GaAs," *J. Electrochem. Soc.* 118:118 (1971).

14. D.W. Shaw, "Enhanced GaAs Etch Rates Near the Edges of a Patterned Mask," *J. Electrochem. Soc.* 113:958 (1966).

15. J.J. Gannon and C.J. Nuese, "A Chemical Etchant for the Selective Removal of GaAs Through SiO_2 Masks," *J. Electrochem. Soc.* 121:1215

16. S. Iida and K. Ito, "Selective Etching of Gallium Arsenide Crystals in the H2SO4-H2O2-H2O System," *J. Electrochem. Soc.* 118:768 (1971).

17. R.A. Logan and F.K. Reinhart, "Optical Waveguides in GaAs-AlGaAs Epitaxial Layers," *J. Appl. Phys.* 44:4172 (1973).

18. J.J. LePore, "Improved Technique for Selective Etching of GaAs and $Ga1_xAlxAs$," *J. Appl. Phys.* 51:6441 (1980).

19. R.P. Tijburg and T van Dongen, "Selective Etching of III-V Compounds with Redox Systems," *J. Electrochem. Soc.* 123: 687 (1976).

20. D.G. Hill, K.L. Lear, and J.S. Harris, "Two Selective Etching Solutions for GaAs on In GaAs and GaAs/AlGaAs on InGaAs," *J. Electrochem. Soc.* 137:2912 (1990).

21. K.E. Bean, "Anisotropic Etching of Si," IEEE Trans. Electron Dev. ED-25:1185 (1978).

22. S. Wolf and R.N. Tauber, "Silicon Processing for the VLSI Era, Vol. 1," Lattice Press, Sunset Beach, CA, (1986).

23. D.L. Kendall, G.R. de Guel, C.D. Fung, P.W. Cheung, W.H. Ko, and D.G. Fleming, "Orientation of the Third Kind: The Coming Age of (110) Silicon Micromachining and Micropackaging of Transducers," Elsevier, Amsterdam, 107 (1985).

24. P.D. Greene, "Selective Etching of Semi-Insulating Gallium Arsenide," Solid-State Electron. 19:815 (1976).

25. H. Muraoka, H.R. Huff and R.R. Burgess, "Controlled Preferential Etching Technology,"Semiconductor Silicon 73:327 (1973).

26. J.C. Greenwood, "Ethylene Diamine-Catechol-Water Mixture Shows Preferential Etching of p-n Junction," *J. Electrochem. Soc.* 116:1325 (1969).

27. D.G. Schimmel, "Dry Etch for (100) Silicon Evaluation," *J. Electrochem. Soc.* 126:479 (1979).

28. F. Secco d'Aragona, "Dislocation Etch for (100) Planes in Silicon," *J. Electrochem. Soc*. 119:948 (1972).

29. E. Sirtl and A. Adler, Z. Metallk, "Chromic acid-hydrofluoric acid as specific reagents for the development of etching pits in silicon," 52:529 (1961).

30. W.C. Dash, "Copper Precipitation on Dislocations in Silicon," *J. Appl. Phys*. 27:1193 (1956).

31. T. Abraham, "GB-288 Chemical Mechanical Polishing Equipment and Materials: A Technical Market Analysis, (2004)", Business Communications Company, Inc., www.bccresearch.com/ advmat/GB288. html (2004).

32. R. Jairath, D. Mukesh, M. Stell, and R. Tolles, "Role of Consumables in the Chemical-Mechanical Polishing (CMP) of Silicon Oxide Films," Proc. 1993 ULSISymp. (1993).

33. M.A. Fury, "Emerging Developments in CMP for Semiconductor Planarization," Solid State Technol. 38(4):47 (1995).

34. G. Nanz and L.E. Camilletti, "Modeling of Chemical-Mechanical Polishing: A Review," IEEE Trans. Semicond. Manuf, 8:382 (1995).

35. F. Malik and M. Hasan, "Manufacturability of the CMP Process," Thin Solid Films, 270:612 (1995).

36. F. Kaufman, S. Cohen, and M. Jaso, "Characterization of Defects Produced in TEOS Thin Films Due to Chemical Mechanical Polishing (CMP)," in *Ultraclean Semiconductor Processing Technology and Surface Chemical Cleaning and Passivation*, MRS 386, Materials Research Society. Pittsburgh, 85 (1995).

37. W.L. Patrick, W.L. Guthrie, C.L. Standley, and P.M. Schiable, "Application of Chemical-Mechanical Polishing to the Fabrication of VLSI Circuit Interconnections," *J. Electrochem. Soc*. 138(6):1778 (1991).

38. B. Davari, C.W. Koburger, R. Schulz, J.D. Warnock, T. Furukawa, M. Jost, Y. Taur, W.G. Schwittek, J.K. DeBrosse, M.L. Kerbaugh, and J.L. Mauer, "*A New Planarization Technique Using a Combination of RIE and Chemical Mechanical Polish* (CMP)," IEDM Tech. Digest, 61 (1989).

39. P. Singer, "Chemical-Mechanical Polishing: A New Focus on Consumables," *Semiconductor Int*. 48 (1994).

40. J.M. Steigerwald, R. Zirpoli, S.P. Murarka, D. Price, and R.J. Gutmann, "Pattern Geometry Effects in the Chemical-Mechanical Polishing of Inlaid Copper Structures," *J. Electrochem. Soc*. 141(10):2842 (1994).

41. R. Capio, J. Farkas, and R. Jairath, "Initial Study on Copper CMP Slurry Chemistries," Thin Solid Films 266(2):238 (1995).

42. E. Ferri, "CMP Chemical Distribution Management," Proc. Semicond. West: Planarization Technology: Chemical Mechanical Polishing (CMP), (1994).

43. F.B. Kaufman, D.B. Thompson, R.E. Broadie, M.A. Jaso, W.L. Gutherie, D.J. Pearson, and M.B. Small, "Chemical-Mechanical Polishing for Fabricating Patterned W Metal Features as Chip Interconnects," *J. Electrochem. Soc.* 138:3460 (1991).

44. C.W. Liu, W.T. Tseng, B.T. Dai, C.Y. Lee, and C.F. Yeh, "Perspectives on the Wear Mechanism During CMP of Tungsten Thin Films," Proc. CMP VLSI/ULSIMultilevel Interconnection Conf., Santa Clara, CA, (1996).

45. I. Kim, K. Murella, J. Schlueter, E. Nikkel, J. Traut, and G. Castleman, "Optimized Process for CMP," *Semicond. Int.*, 9:119 (1996).

46. R. Capio, J. Farkas, and R. Jairath, "Initial Study on Copper CMP Slurry Chemistries," Thin Solid Films 266:238 (1995).

47. C. Sainio, D. Duquette, J. Steigerwald, and S. Muraka, "Electrochemical Effects in the Chemical-Mechanical Polishing of Copper for Integrated Circuits," *J. Elect. Mater.* 25(10):1593 (1996).

48. I. Ali, S.R. Roy, and G. Shinn, "Chemical-Mechanical Polishing of Interlayer Dielectric: A Review," Solid State Technol. 37(10):63 (1994).

49. L. Borucki Leonard J. Borucki, T. Witelski, C. Please, P.R. Kramer and D. Schwendeman, "A Theory of Pad Conditioning for Chemical-Mechanical Polishing," *J. Eng. Math.* 50(1):1-24 (2004).

50. J.M. de Larios, M. Ravkin, D.L. Hetherington, and J.D. Doyle, "Post-CMP Cleaning for Oxide and Tungsten Applications," Semicond. Int. 121 (1996).

51. I. Malik, J. Zhang, A.J. Jensen, J.J. Farber, W.C. Krusell, S. Raghavan, and Rajhunath, "Post-CMP Cleaning of W and SiO_2: A Model Study," *in Ultraclean Semiconductor Processing Technology and Surface Chemical Cleaning and Passivation,* MRS 386, Materials Research Society, Pittsburgh, 109 (1995).

52. T.J. Cotler and M. Elta, "Plasma-Etch Technology," IEEE Circuits, Dev. Mag. 6:38 (1990).

53. D.M. Manos and D.L. Flamm, "Plasma Etching, An Introduction," Academic Press, Boston, (1989).

54. R.A. Morgan, "Plasma Etching in Semiconductor Fabrication," Elsevier, Amsterdam, (1985).

55. A.J. van Roosmalen, J.A.G. Baggerman, and S.J.H. Brader, "Dry Etching for VLSI," Plenum, New York, (1991).

56. J.W. Coburn and H.F. Winters, "Ion and Electron Assisted Gas Surface Chemistry," *J. Appl. Phys.* 50(5): 3189 to 3196 (1979).

57. V.M. Donelly, D.I. Flamm, W.C. Dautremont-Smith, and D.J. Werder, "Anisotropic Etching of SiO_2 in Low-Frequency CF_4/O_2 and NF_3/Ar Plasmas," *J. Appl. Phys.* 55:242 (1984).

58. G. Smolinsky and D.L. Flamm, "The Plasma Oxidation of CF4 in a Tubular, Alumina, Fast-Flow Reactor," *J. Appl. Phys.* 50:4982 (1979).

59. C.J. Mogab, A.C. Adams, and D.L. Flamm, "Plasma Etching of Si and SiO_2—The Effect of Oxygen Additions to CF_4 Plasmas," *J. Appl. Phys.* 49:3796 (1978).

60. J.W. Coburn, "In-situ Auger Spectroscopy of Si and SiO_2 Surfaces Plasma Etched in CF_4-H_2 Glow Discharges," *J. Appl. Phys.* 50:5210 (1979).

61. R. d'Agostino, F. Cramarossa, F. Fracassi, E. Desimoni, L. Sabbatini, P.G. Zambonin, and G. Caporiccio, "Polymer Film Formation in C_2F_6-H_2 Discharges," *Thin Solid Films* 143:163 (1986).

62. M. Shima, "A Study of Dry-Etching Related Contaminations of Si and SiO_2," *Surf Sci.* 86:858 (1979).

63. S. Joyce, J.G. Langan, and J.I. Steinfeld, "Chemisorption of Fluorocarbon Free Radicals on Si and SiO_2," J. Chem. Phys. 88:2027 (1988).

64. C. Cardinaud and G. Turban, "Mechanistic Studies of the Initial Stages of Si and SiO_2 in a CHF_3 Plasma," *Appl. Surf. Sci.* 45:109 (1990).

65. G.S. Oehrlein, K.K. Chan, and G.W. Rubloff, "Surface Analysis of Realistic Semiconductor Microstructures," *J. Vacuum. Sci. Technol. A* 7:1030 (1989).

66. G.S. Oehrlein and J.F. Rembetski, "Study of Sidewall Passivation and Microscopic Silicon Roughness Phenomena in Chlorine-based Reactive Ion Etching of Silicon Trenches,"*J. Vacuum Sci. Technol. B* 8:1199 (1990).

67. K.V. Guinn and C.C. Chang, "Quantitative Chemical Topography of Polycrystalline Si Anisotropically Etched in Cl_2/O_2 High Density Plasmas," *J. Vacuum Sci. Technol. B* 13:214 (1995).

68. K.V. Guinn and V.M. Donnelly, "Chemical Topography of Anisotropic Etching of Polycrystalline Si Masked with Photoresist," *J. Appl. Phys.* 75:2227 (1994).

69. F.H. Bell and O. Joubert, "Polycrystalline Gate Etching in High Density Plasmas. II. X-Ray Photoelectron Spectroscopy Investigation of Silicon Trenches Etched Using a Chlorine-based Chemistry," *J. Vacuum Sci. Technol.* B 14:1796 (1996).
70. M.M. Millard and E. Kay, "Difluocarbene Emission Spectra from Fluorocarbon Plasmas and Its Relationship to Fluorocarbon Polymer Formation," *J. Electrochem. Soc.* 129:160 (1982).
71. J.W. Coburn and E. Kay, "Some Chemical Aspects of Fluorocarbon Plasma Etching of Silicon and Its Compounds," *IBM J. Res. Dev.* 23:33 (1979).
72. V.M. Donnelly, D.E. Ibbotson, and D.L. Flamm, "Ion Bombardment Modification of Surfaces: Fundamentals and Applications," Elsevier, New York, (1984).
73. F.H.M. Sanders, J. Dieleman, H.J.B. Peters, and J.A. M. Sanders, "Selective Isotropic Dry Etching of Si_3N_4 over SiO_2," *J. Electrochem. Soc.* 129:2559 (1982).
74. C.J. Mogab, "The Loading Effect in Plasma Etching," *J. Electrochem. Soc.* 124:1262 (1977).
75. B.A. Heath and T.M. Mayer, "VLSI Electronics Microstructure Science Plasma Processing for VLSI," Academic Press, New York, (1984).
76. J.M.E. Harper, "Ion Beam Techniques in Thin Film Deposition," *Solid State Technol.* 30:129 (1987).
77. R.E. Lee, "Ion-Beam Etching (Milling)," VLSI Electronics Microstructure Science 8, Plasma Processing for VLSI, Academic Press, New York, (1984).
78. H.R. Kaufman, J.J. Cuomo, and J.M.E. Harper, "Techniques and Applications of Broad-Beam Ion Sources Used in Sputtering—Part 1. Ion Source Technology," *J. Vacuum Sci. Technol.* 21:725 (1982).
79. J.M.E. Harper, "Ion Beam Etching-Plasma Etching: An Introduction," Academic Press, New York, (1989).
80. S. Matsup and Y. Adachi, "Reactive Ion Beam Etching Using a Broad Beam ECR Ion Source," *Jpn. J. Appl. Phys.* 21:L4 (1982).

6

Diffusion

6.1 INTRODUCTION

The selectively changing of electrical properties of silicon through the introduction of impurities commonly referred to as dopants. In the early years of integrated circuit fabrication, deep semiconductor junctions required doping processes followed by a drive-in "step to diffuse the dopants to the desired depth", i.e. diffusion was required to successfully fabricated devices. Impurity atoms are introduced onto the surface of a silicon wafer and diffuse into the lattice because of their tendency to move from regions of high to low concentration. The doping concentration decreases monotonically from the surface, and the in-depth distribution of the dopant is determined mainly by the temperature and diffusion time. Diffusion of impurity atoms into silicon crystal takes place only at elevated temperature, typically 900 to 1200°C.

Diffusion and ion implantation are the two key processes to introduce a controlled amount of dopants into semiconductors and to alter the conductivity type. In modern state-of-the-art IC fabrication the required junction depths have become so shallow that dopants are introduced into the silicon at the desired depth by ion implantation and any diffusion of the dopants is unwanted; therefore diffusion has become a problem as opposed to an asset. There are many non-state-of-the-art processes still in use throughout that industry where doping and diffusion are still in use and for state-of-the-art processes diffusion must be understood in order to minimize undesired effects. Generally speaking, diffusion and ion implantation complement each other. For instance, diffusion is used to form a deep junction, such as an n-tub in a CMOS device, while ion implantation is utilized to form a shallow junction, like a source / drain junction of a MOSFET.

Boron is the most common p-type impurity in silicon, whereas arsenic and phosphorus are used extensively as n-type dopants. These three elements are highly soluble in silicon with solubilities exceeding >10^{20} atoms/cm^3 in the diffusion temperature range (between 80°C and 120°C). These dopants can be introduced via several means, including solid sources (BN for B, As_2O_3 for As, and P_2O_5 for P), liquid sources (BBr_3, $AsCl_3$, and $POCl_3$), and gaseous sources (B_2H_6, AsH_3, and PH_3). Usually, the gaseous source is transported to the semiconductor surface by an inert gas (e.g. N_2) and is then reduced at the surface. After the impurities are introduced they may redistribute in the wafer. This may be intentional or it may be a parasitic effect of a thermal process. In either event, it must be controlled and monitored. The motion of impurity atoms in the wafer occurs primarily by diffusion, the net movement of a material that occurs near a concentration gradient as a result of random thermal motion.

6.2 ATOMIC MECHANISMS OF DIFFUSION

Impurity atoms utilized as dopants such as boron (B), phosphorus (P) and arsenic (As) occupy substitution positions where the dopant atoms can contribute free electrons or holes to the silicon lattice (Dopant atoms introduced to silicon by ion implantation may not occupy substitution positions until the dopant is activated). The diffusion of impurities into a solid is basically the same type of process as occurs when excess carriers are created non-uniformly in a semiconductor which cause carrier gradient. In each case, the diffusion is a result of random motion, and particles diffuse in the direction of decreasing concentration gradient. The random motion of impurity atoms in a solid is, of course, rather limited unless the temperature is high.

There are mainly two types of physical mechanisms by which the impurities can diffuse into the lattice. They are

6.2.1 Substitutional Diffusion

At high temperature many atoms in the semiconductor move out of their lattice site, leaving vacancies into which impurity atoms can move. The impurities, thus, diffuse by this type of vacancy motion and occupy lattice position in the crystal after it is cooled. Thus, substitution diffusion takes place by replacing the silicon atoms of parent crystal by impurity atom. In other words, impurity atoms diffuse by moving from a lattice site to a neighboring one by substituting

6.2 Atomic mechanisms of Diffusion

for a silicon atom which has vacated a usually occupied site as shown in the figure 6.1 below.

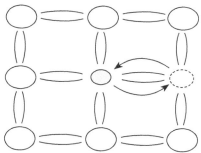

Fig. 6.1 Substitutional Diffusion

Substitutional diffusion mechanism is applicable to the most common diffusants, such as B, P and As. These dopants atoms are too huge to fit into the interstices or voids, so the only way they can enter the silicon crystal is to substitute for a Si atom.

In order for such an impurity atom to move to a neighboring vacant site, it has to overcome energy barrier which is due to the breaking of covalent bonds. The probability of its having enough thermal energy to do this is proportional to an exponential function of temperature. Also, whether it is able to move is also dependent on the availability of a vacant neighboring site and since an adjacent site is vacated by a Si atom due to thermal fluctuation of the lattice, the probability of such an event is again an exponent of temperature.

The jump rate of impurity atoms at ordinary temperatures is very slow, for example about 1 jump per 10^{50} years at room temperature. However, the diffusion rate can be speeded up by an increase in temperature. At a temperature of the order 1000°C, substitutional diffusion of impurities is practically realized in sensible time scales.

6.2.2 Interstitial Diffusion

In such, diffusion type, the impurity atom does not replace the silicon atom, but instead moves into the interstitial voids in the lattice. The main types of impurities diffusing by such mechanism are gold, copper, and nickel. Gold, particularly, is introduced into silicon to reduce carrier life time and hence useful to increase speed at digital IC's.

Because of the large size of such metal atoms, they do not usually substitute in the silicon lattice. To understand interstitial diffusion, let us consider a unit cell of the diamond lattice of the silicon which has five interstitial voids. Each of the voids is big enough to contain an impurity atom. An impurity atom located in one such void can move to a neighboring void, as shown in the figure 6.2 below.

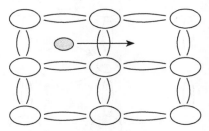

Fig. 6.2 Interstitial Diffusion

In doing so it again has to surmount a potential barrier due to lattice this time, most neighboring interstitial sites are vacant so the frequency of movement is reduced. Again, the diffusion rate due to this process is very slow at room temperature but becomes practically acceptable at normal operating temperature of around 1000°C. It will be noticed that the diffusion rate due to interstitial movement is much greater than for substitutional movement. This is possible because interstitial diffusants can fit in the voids between silicon atoms. For example, lithium acts as a donor impurity in silicon, it is not normally used because it will still move around even at temperatures near room temperature, and thus will not be frozen in place. This is true of most other interstitial diffusions, so long-term device stability cannot be assured with this type of impurity.

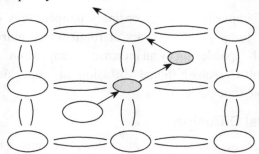

Fig 6.3 Interstitialcy diffusion

Substitutional diffusion happens when a substitutional atom exchanges lattice positions with a vacancy-requires the presence of a vacancy. Interstitial diffusion occurs when an interstitial atom jumps to another interstitial position. Interstitialcy diffusion results from silicon self-interstitials displacing substitutional impurities to an interstitial position-requires the presence of silicon self-interstitials, the impurity interstitial may then knock a silicon lattice atom into a self-interstitial position. It is important to observe here that since dopant atoms such as P, As and B occupancy substitutional positions once activated, dopant diffusion is closely linked to and controlled by the presence of vacancy and interstitial point defects.

6.3 FICK'S LAWS OF DIFFUSION

The diffusion rate of impurities into semiconductor lattice depends on the following:

- Mechanism of diffusion
- Temperature
- Physical properties of impurity
- The properties of the lattice environment
- The concentration gradient of impurities
- The geometry of the parent semiconductor

The behavior of diffusion particles is governed by Fick's Law, which when solved for appropriate boundary conditions, gives rise to various dopant distributions, called profiles which are approximated during actual diffusion processes.

In 1855, Fick drew analogy between material transfer in a solution and heat transfer by conduction. Fick assumed that in a dilute liquid or gaseous solution, in the absence of convection, the transfer of solute atoms per unit area in a one-dimensional flow can be described by the following equation.

$$F = -D \frac{\partial N(x,t)}{\partial x} = -\frac{\partial F(x,t)}{\partial x} \quad \ldots (6.1)$$

Where F is the rate of transfer of solute atoms per unit area of the diffusion flux density (atoms/cm^2-sec), N is the concentration of solute atoms (number of atoms per unit volume/cm^3), and x is the direction of solute flow. (Here N is assumed to be a function of x and t only), t is the diffusion time, and D is the diffusion constant (also referred to as diffusion coefficient or diffusivity) and has units of cm^2/sec.

The above equation is called Fick's First law of diffusion and states that the local rate of transfer (local diffusion rate) of solute per unit area per unit time is proportional to the concentration gradient of the solute, and defines the proportionality constant as the diffusion constant of the solute. The negative sign appears due to opposite direction of matter flow and concentration gradient. That is, the matter flows in the direction of decreasing solute concentration.

Fick's first law is applicable to dopant impurities used in silicon. In general the dopant impurities are not charged, nor do they move in an electric field, so the usual drift mobility term (as applied to electrons and holes under the influence of electric field) associated with the above equation can be omitted. In this equation N is in general function of x, y, z and t.

The change of solute concentration with time must be the same as the local decrease of the diffusion flux, in the absence of a source or a sink. This follows from the law of conservation of matter. Therefore we can write down the following equation

$$\frac{\partial N(x,t)}{\partial t} = -\frac{\partial F(x,t)}{\partial x} \qquad \ldots (6.2)$$

Substituting 'F' value in the above equation:

$$\frac{\partial N(x,t)}{\partial x} = \frac{\partial}{\partial x}\left[D * \frac{\partial N(x,t)}{\partial x}\right] \qquad \ldots (6.3)$$

The diffusion constant can be considered as a constant if the concentration of the solute is low at a given temperature.

So the equation becomes

$$\frac{\partial N(x,t)}{\partial x} = D\left[\frac{\partial^2 N(x,t)}{\partial^2 x}\right] \qquad \ldots (6.4)$$

This is Fick's second law of distribution. Fick's second law is identical in form to the equation for heat conduction differing only in the constant D, and therefore the large body of work on heat flow can be applied to the problems of impurity atom diffusion in silicon.

6.4 DIFFUSION PROFILES

The diffusion profile of dopant atoms is dependent on the initial and boundary conditions. Depending on boundary equations the Fick's Law has two types of solutions. These solutions provide two types of impurity distribution namely constant source distribution following complimentary error function (erfc) and limited source distribution following Gaussian distribution function. Solutions for **equation 6.3** have been obtained for various simple conditions, including constant-surface-concentration diffusion and constant-total-dopant diffusion. In the first scenario, impurity atoms are transported from a vapor source onto the semiconductor surface and diffuse into the semiconductor wafer. The vapor source maintains a constant level of surface concentration during the entire

6.4 Diffusion Profiles

diffusion period. In the second situation, a fixed amount of dopant is deposited onto the semiconductor surface and is subsequently diffused into the wafer.

6.4.1 Constant Source Concentration Distribution

In this impurity distribution, the impurity concentration at the semiconductor surface is maintained at a constant level throughout the diffusion cycle i.e.

$$N(o, t) = N_S = \text{Constant}$$

The solution to the diffusion equation which is applicable in this situation is most easily obtained by first considering diffusion inside a material in which the initial concentration changes in same plane as x=0, from N_S to 0. The initial condition at t = 0 is N(x, 0) = 0 which states that the dopant concentration in the host semiconductor is initially zero. The boundary conditions are:

$$N(o, t) = N_S = \text{Constant and}$$

Where N_s is the surface concentration (at x = 0) that is independent of time. The second boundary condition states that at large distances from the surface, there are no impurity atoms. The solution of the differential equation that satisfies the initial and boundary conditions is given by.

$$N(x, t) = N_s erfc\left\{\frac{x}{2\sqrt{Dt}}\right\} \quad \ldots (6.5)$$

Where erfc stands for the complementary error function, x is the distance, D is the diffusion coefficient, t is the diffusion time and is the diffusion length,. Few erfc function properties are summarized in Table 6.1.

Table 6.1 Error Function properties

g	erfc (y)
0	1.0
0.5	0.5
1.0	1.7
1.5	0.35
2.0	0.005
2.1	0.0004
∞	0

The constant surface concentration condition diffusion profile is exhibited in Figure 6.4 for both linear and logarithmic scales. The total number of dopants per unit area of the semiconductor, Q(t), is given by integrating N(x, t) from x = 0 to x = ∞:

$$Q(t) = \frac{2}{\sqrt{\pi}} N_s \sqrt{Dt} \cong (1.13) N_s \sqrt{Dt} \quad \ldots (6.6)$$

The gradient of the diffusion (dx/dN) can be found by differentiating Equation 6.5, and the result is:

$$\frac{dN}{dx} = -\frac{N_s}{\sqrt{\pi Dt}} \exp\left\{\frac{-x^2}{4Dt}\right\} \quad \ldots (6.7)$$

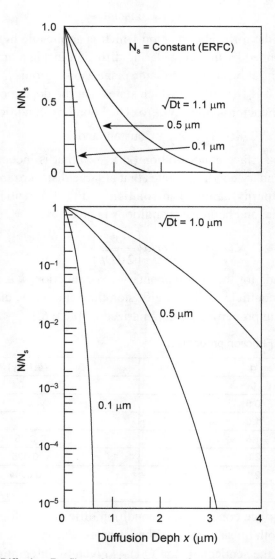

Fig. 6.4 (a) Diffusion Profiles of erf vs distance for successive diffusion times

6.4 Diffusion Profiles

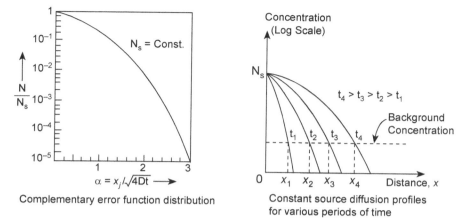

Complementary error function distribution

Constant source diffusion profiles for various periods of time

Fig. 6.4 (b) erf profile evaluation vs time

If the diffused impurity type is different from the resistivity type of the substrate material, a junction is formed at the points where the diffused impurity concentration is equal to the background concentration already present in the substrate.

In the fabrication of monolithic IC's, constant source diffusion is commonly used for the isolation and the emitter diffusion because it maintains a high surface concentration by a continuous introduction of dopant. There is an upper limit to the concentration of any impurity that can be accommodated at the semiconductor wafer at some temperature. This maximum concentration which determines the surface concentration in constant source diffusion is called the solid solubility of the impurity.

6.4.2 Limited Source Diffusion or Gaussian Diffusion

Here a determined amount of impurity is introduced into the crystal unlike constant source diffusion The diffusion takes place in two steps:

1. **Predeposition Step** – In this step a fixed number of impurity atoms are deposited on the silicon wafer durings short time.
2. **Drive-in step** – Here the impurity source is turned off and the amounts of impurities already deposited during the first step are allowed to diffuse into silicon water.

For this case, a fixed (or constant) amount of dopant is deposited onto the semiconductor surface in a thin layer, and the dopant is subsequently diffused into the semiconductor. The initial condition at $t = 0$ is again $(x, 0) = 0$. The boundary conditions are:

$\int_0^\infty N(x,t) = S$ and $N(\infty,t) = (x, 0) = 0$. The boundary conditions are:

$$\int_0^\infty N(x,t) = S \text{ and } N(\infty, t) = 0 \qquad \ldots (6.8)$$

where S is the total amount of dopant per unit area, the solution of the diffusion equation satisfying the above conditions is

$$N(x, t) = \frac{S}{\sqrt{\pi Dt}} \exp\left\{\frac{x^2}{4Dt}\right\} \qquad \ldots (6.9)$$

This expression is the Gaussian distribution, and the dopant profile is displayed in figure 6.5. By substituting $x = 0$ into equation 6.9:

$$N_s(t) = \frac{S}{\sqrt{\pi Dt}} \qquad \ldots (6.10)$$

The dopant surface concentration therefore decreases with time, since the dopant will move into the semiconductor as time increases. The gradient of the diffusion profile is obtained by differentiating equation 6.9:

$$\frac{dN}{dx} = -\frac{x}{2Dt} N(x,t) \qquad \ldots (6.11)$$

The gradient is zero at $x = 0$ and $x = \infty$, and the maximum gradient occurs $x = \sqrt{2Dt}$.

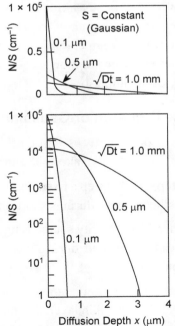

Fig. 6.5. Diffusion profile of Normalized Gaussian Function vs distance for successive times

6.5 Dual Diffusion Process

Both the complementary error function and the Gaussian distribution are function of a normalized distance. Hence, if we normalize the dopant concentration with the surface concentration, each distribution can be represented by a single curve valid for all diffusion times, as shown in figure 6.6. The essential difference between the two types of diffusion techniques is that the surface concentration is held constant for error function diffusion. It decays with time for the Gaussian type owing to a fixed available doping concentration (S). For the case of modeling the depletion layer of a p-n junction, the erfc is modeled as a step junction and the Gaussian as a linear graded junction. In the case of the erfc, the surface concentration is constant, typically the maximum solute concentration at that temperature or solid solubility limit.

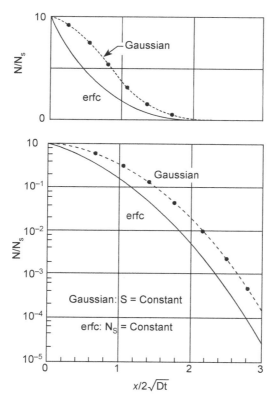

Fig. 6.6 Doping profile for normalized concentration vs normalized distance of erf and Gaussian function

6.5 DUAL DIFFUSION PROCESS

In VLSI processing, a two-step diffusion sequence is commonly used, in which a predeposition diffusion layer is formed under a constant-surface-concentration condition and is followed by a drive-in diffusion or redistribution under a constant- total-dopant condition. For most practical cases, the diffusion length $\sqrt{Dt}$ for the predeposition diffusion is much smaller than that for the drive-in condition. Hence, the predeposition profile can be treated as a delta function at the surface.

Parameters which Affect Diffusion Profile

- **Solid Solubility** – In deciding which of the availability impurities can be used, it is essential to know if the number of atoms per unit volume required by the specific profile is less than the diffusant solid solubility.
- **Diffusion temperature** – Higher temperatures give more thermal energy and thus higher velocities, to the diffused impurities. It is found that the diffusion coefficient critically depends upon temperature. Therefore, the temperature profile of diffusion furnace must have higher tolerance of temperature variation over its entire area.
- **Diffusion time** – Increases of diffusion time, t, or diffusion coefficient D have similar effects on junction depth as can be seen from the equations of limited and constant source diffusions. For Gaussian distribution, the net concentration will decrease due to impurity compensation, and can approach zero with increasing diffusion tunes. For constant source diffusion, the net Impurity concentration on the diffused side of the p-n junction shows a steady increase with time.
- **Surface cleanliness and defects in silicon crystal** – The silicon surface must be prevented against contaminants during diffusion which may interfere seriously with the uniformity of the diffusion profile. The crystal defects such as dislocation or stacking faults may produce localized impurity concentration. This results in the degradation of junction characteristics. Hence silicon crystal must be highly perfect.

Basic Properties of the Diffusion Process

Following properties could be considered for designing and laying out ICs.

- When calculating the total effective diffusion time for given impurity profile, one must consider the effects of subsequent diffusion cycles.
- The erfc and gaussian functions show that the diffusion profiles are functions of $\frac{x}{\sqrt{Dt}}$. Hence, for a given surface and background concentration, the junction depth x_1 and x_2 associated with the two separate diffusions having different times and temperature.

6.5 Dual Diffusion Process

- **Lateral Diffusion Effects** – The diffusions proceed sideways from a diffusion window as well as downward. In both types of distribution function, the side diffusion is about 75 to 80 per cent of the vertical diffusion.

6.5.1 Intrinsic & Extrinsic Diffusion

Diffusion that occurs when the doping concentration is lower than the intrinsic carrier concentration (n_i), at the diffusion temperature is called intrinsic diffusion. In this region, the resulting dopant profiles of sequential or simultaneous diffusion of n-type or p-type impurities can be determined by superposition, that is, the diffusion processes can be treated independently.

However, when the dopant concentration exceeds n_i (e.g. at 1000°C, $n_i = 5 \times 10^{18}$ atoms/cm³), the process becomes extrinsic, and the diffusion coefficients become concentration dependent, as shown in figure 6.7.

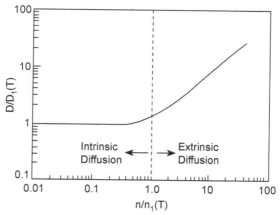

Fig. 6.7 Donor impurity diffusion coefficient vs electron concentration

In order to make diffusion calculations of reasonable accuracy Fair's diffusivity model will be presented in the next few sections. Fair has developed a model for diffusivity based on impurity interactions with charged vacancy states that provides a reasonably good fit to most observed diffusion results. Fair's model calculates diffusivity as the sum of the individual charged vacancy - impurity diffusivities.

The intrinsic model is given by

$$D_i = D^0 + D^+ + D^- + D^= + \ldots \quad \text{for n or p} \ll n_i \qquad \ldots (6.12)$$

where: D_i is the intrinsic diffusivity D^0 is the neutral vacancy-impurity diffusion, D^+ is the positively charged vacancy-impurity diffusion, D^- is the negatively charged vacancy-impurity diffusion and $D^=$ is the doubly negatively charged vacancy - impurity diffusivity and n_i is the intrinsic carrier concentration For extrinsic silicon

$$D_x = D^0 + D^+\left[\frac{p}{n_i}\right] + D^-\left[\frac{n}{n_i}\right] + D^=\left[\frac{n}{n_i}\right]^2 \quad \text{for n or p} >> n_i \quad ...(6.13)$$

where, D_x is the extrinsic diffusivity. For unambiguous impurities not all vacancy charge state - impurity combinations will participate in the diffusivity.

In order for an impurity to diffuse done silicon, the impurity must either move around silicon atoms or displace silicon atoms. During interstitial diffusion the diffusing atom jumps from one interstitial position to another interstitial position, with relatively low barrier energy and a relatively high number of interstitial sites. Substitutional atoms need the presence of a vacancy or an interstitial to diffuse break lattice bonds. Vacancy and interstitial formation are relatively high-energy processes and so are relatively rare in equilibrium. Breaking bonds to the lattice is also a relatively high-energy process and substitutional atoms tend to diffuse at a much lower rate than interstitial atoms. When a host atom acquires sufficient energy and leaves its lattice site, a vacancy is created. Depending on the charges associated with the vacancy, it can have:

(1) A neutral vacancy, V^o,

(2) An acceptor vacancy, V^-,

(3) A doubly-charged acceptor vacancy $V^=$,

(4) A donor vacancy, V^+,

(5) and others.

The vacancy density of a given charge state (i.e. the number of vacancies per unit volume) has temperature dependence similar to that of the carrier density:

$$N_v = N_i e^{\frac{E_F - E_i}{kT}} \quad ...(6.14)$$

where N_v is the vacancy density, N_i is the intrinsic vacancy density, E_F is the Fermi level, and E_i is the intrinsic Fermi level.

If the dopant diffusion is dominated by the vacancy mechanism, the diffusion coefficient is estimated to be proportional to the vacancy density. At low doping concentrations ($n<n_i$), the Fermi level coincides with the intrinsic Fermi level (i.e., $E_F = E_i$). The vacancy density is equal to N_i and independent of the dopant concentration. The diffusion coefficient, which is proportional to N_i, will also be independent of doping concentration. At high doping concentrations ($n>n_i$),

6.5 Dual Diffusion Process

the Fermi level will move toward the conduction band edge for donor-type vacancies, and the term becomes larger than unity. This causes N_v to increase, which in turn gives rise to enhanced diffusion, as exhibited in figure 6.8.

Even though it is now known that substitutional diffusion is not strictly vacancy dominated, Fair's model is useful for the reasonable results it obtains without resorting to complex simulation. The temperature dependence of the diffusivity values presented in the next several sections will take the arhenius form

$$D = D_0 e^{-E_a/kT} \quad \ldots(6.15)$$

Where D_0 is a pre exponential constant, Arhenius equations form a straight line when ln(D) is plotted versus 1/T.

6.5.2 Diffusivity of Antimony in Silicon

Antimony is believed to diffuse by a purely vacancy mechanism with the Sb^+V^- dominating. The intrinsic diffusivity of antimony is given by $D_i = 0.214 e^{-3.65/kT}$ cm²/s

The extrinsic diffusivity of antimony is given by

$$D_x = D^0 + D^-\left[\frac{n}{n_i}\right] \quad \ldots(6.16)$$

here D^+ and $D^= = 0$,

D^0 and D^- are given by $D^0 = 0.214 e^{-3.65/kT}$ cm²/s and $D^- = 15.0 e^{-4.08/kT}$ cm²/s

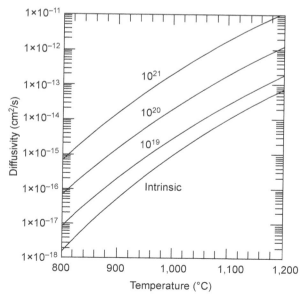

Fig. 6.8 Diffusivity of antimony in Silicon vs temperature and concentration

The diffusivity of antimony versus temperature is illustrated in figure 6.8. Antimony has a large tetrahedral radius of 0.136 nms versus a radius of 0.118 nms for silicon giving a 1.15 mismatch factor and creating strain in the silicon lattice. Antimony diffusion is retarded by oxidation due to injected silicon self-interstitials annihilating vacancies.

6.5.3 Diffusivity of Arsenic in Silicon

Arsenic is believed to diffuse primarily through a vacancy mechanism with an interstitialcy component. The arsenic interstitial formation energy is estimated to be a relatively high 2.5 eV. Above approximately 1, 0.50°C diffusion is dominated by AsV^- vacancy pairs with $As^+V^=$ being moderately rare due to the small pair binding energy. Oxidation has little effect on arsenic diffusivity due to the high arsenic interstitial formation energy.

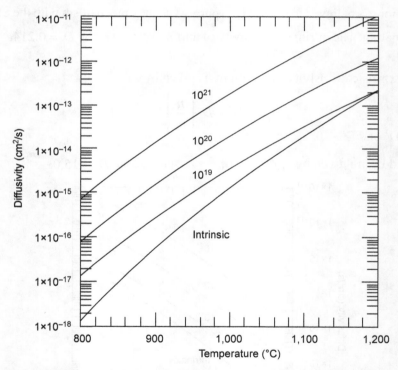

Fig. 6.9 Diffusivity of arsenic in Silicon vs temperature and concentration

The intrinsic diffusivity of arsenic is given by

$$D_i = 22.9 e^{-4.1/kT} \qquad \ldots (6.17)$$

At high concentration ($>10^{20}$ atoms/cm^3) arsenic diffusivity is complicated by clustering. Arsenic atoms are believed to form clusters where three arsenic

6.5 Dual Diffusion Process

atoms bond with an electron. Clusters are virtually immobile at T<1000°C. Above ~1000°C arsenic un-clusters and equations below again gives the diffusivity.

$$D_x = D^0 + D^- \left[\frac{n}{n_i}\right] \quad \ldots (6.18)$$

and D^+ and $D^= = 0$, D^0 and D^- are given by

$$D^0 = 0.066 e^{-3.44/kT} \text{ cm}^2/\text{s}$$

and

$$D^- = 12.0 e^{-4.05/kT} \text{ cm}^2/\text{s}$$

Diffusivity of arsenic versus doping is plotted in **figure 6.9**

Arsenic has exactly the same tetrahedral diameter as silicon and so arsenic does not strain the silicon lattice or induce enhanced diffusivity in other dopants for T~700°C.

6.5.4 Diffusivity of Boron in Silicon

In Fair's model of diffusion, boron is assumed to diffuse exclusively by a vacancy mechanism under non-oxidizing conditions. Estimates of the interstitialcy component of boron diffusion are presented in table 6.2 and have recently been estimated at < 98%.

Table 6.2 Interstitialcy component of boron diffusion

Temperature (°C)			
950°C	1,000°C	1050°C	1100°C
0.16	0.17	0.18	0.19
0.22	0.34	0.42	0.52
0.17	0.32	0.17	0.17
0.38	0.17	0.60	0.80

In Fair's model boron diffuses by a $B^- V^-$ vacancy pair with migration energy approximately 0.5 eV lower than other vacancy-ion pairs. Boron diffusivity is enhanced by p dopants when $p > n_i$ and reduced for $p < n_i$, boron diffusion is actually retarded in N-type silicon where $n > n_i$.

The intrinsic diffusivity of boron is given by

$$D_i = 0.76 \, e^{-3.46/kT} \text{ cm}^2/\text{s} \quad \ldots (6.19)$$

The extrinsic diffusivity of antimony is given by

$$D_x = D^0 + D^+ \left[\frac{p}{n_i}\right] \quad \ldots (6.20)$$

and D^- and $D^= = 0$, D^0 and D^+ are given by

$$D^0 = 0.037 e^{-3.46/kT} \text{ cm}^2/\text{s}$$

and
$$D^+ = 0.76 e^{-3.46/kT} \text{ cm}^2/\text{s}$$

The diffusivity of boron versus doping is plotted in figure 6.10

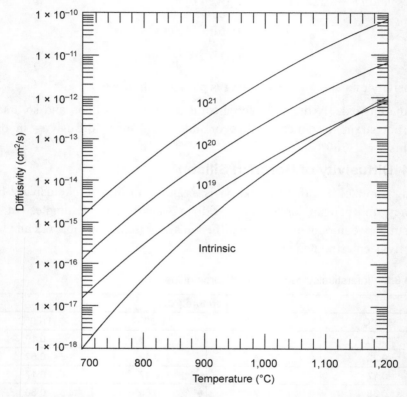

Fig. 6.10 Diffusivity of boron in Silicon vs temperature and concentration

Boron is a faster diffuser than either phosphorus or arsenic. Boron has a tetrahedral radius of 0.82 nms versus 1.18 nms for silicon or a 0.75 mismatch ratio. The relatively large size mismatch for boron versus silicon produces lattice strain that can lead to dislocation formation and reduced diffusivity. The strain introduced by boron into the silicon lattice can help to getter impurities by a trapping mechanism. High concentrations of boron are particularly good at guttering iron. Boron has a relatively low energy for interstitial formation of 2.26 eV, therefore boron diffusivity is enhanced by oxidation.

6.5 Dual Diffusion Process

6.5.5 Diffusivity of Phosphorus in Silicon

The diffusivity of phosphorus is again explained as a vacancy dominated diffusion. Phosphorus diffusion exhibits three distinct regions of behavior illustrated in figure 6.11.

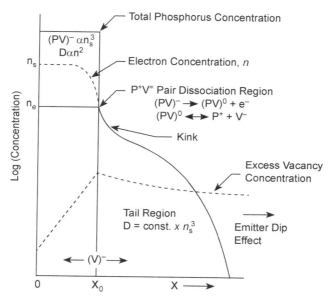

Fig. 6.11 Phosphrous diffusion profile and vacancy model

The phosphorus profile has three distinct regions:
1. The high concentration region where the total phosphorus concentration exceeds the free carrier concentration.
2. A kink in the profile
3. A tail region of enhanced diffusivity

In the high concentration region a fraction of P^+ ions pair with $V^=$ vacancies as $P^+V^=$ pairs noted as $(PV)^-$. The concentration of $(PV)^-$ pairs is proportional to the surface electron concentration cubed (n_s^3), which has to be determined experimentally.

The intrinsic diffusivity of phosphorus is given by

$$D_i = 3.85 \, e^{-3.66/kT} \text{ cm}^2/\text{s} \qquad \ldots (6.21)$$

The extrinsic diffusivity of antimony is given by

$$D_x = D^0 + D^= \left[\frac{n}{n_i}\right]^2 \qquad \ldots (6.22)$$

and $D^- \sim 0$ and $D^+ = 0$, D^0 and $D^=$ are given by

$$D^0 = 3.85\, e^{-3.66/kT}\, \text{cm}^2/\text{s}$$

and
$$D^= = 42.2\, e^{-4.37/kT}\, \text{cm}^2/\text{s}$$

Phosphorus has a tetrahedral radius of 0.110 nms versus 0.118 nms for silicon resulting in a mismatch ratio of 0.93, and at high concentrations – phosphorus lattice strain can lead to defect formation. Fair and Tsai have proposed that tail formation in the phosphorus profile is due to dissociation of $P^+V^=$ pairs when n drops below $10^{20}/\text{cm}^3$ at the diffusion front (the Fermi level is ~0.11 eV from the conduction band), the $V^=$ vacancy changes state and the binding energy decreases enhancing the probability of disassociation and increasing the vacancy flux. The diffusivity of phosphrous versus doping is plotted in figure 6.12.

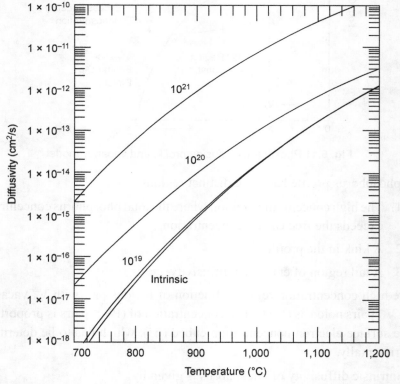

Fig. 6.12 Diffusivity of phosphrous in Silicon vs temperature and concentration

6.6 EMITTER PUSH EFFECT

In silicon n-p-n bipolar transistors employing a phosphorus-diffused emitter and a boron-diffused base, the base region under the emitter region (inner base) is deeper by up to 0.6 µm than that outside the emitter region (outer

6.6 Emitter Push Effect

based). This phenomenon is called the emitter push effect, as illustrated in Figure 6.13. The dissociation of phosphorus vacancy ($P^+V^=$) pairs at the kink region provides a mechanism for the enhanced diffusion of phosphorus in the tail region. The diffusivity of boron under the emitter region (inner base) is also enhanced by the dissociation of P^+V pairs.

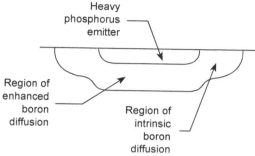

Fig. 6.13 Enhanced diffusivity under heavy phosphorous doped regions – Emitter Push

Diffusivity of Miscellaneous Dopants in Silicon

In the preceding sections the diffusivity of the four most common silicon dopants was discussed in detail. There are a variety of other impurities whose

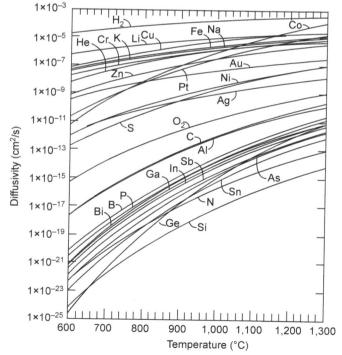

Fig. 6.14 Diffusivity of selected impurities in silicon vs temperature

diffusivity may be of interest. The size mismatch between a diffusing specie and the silicon lattice can result in lattice strain which may enhance the diffusivity of co-diffusing specie figure 6.14 represents relative diffusivity of different elements in silicon

6.7 FIELD-AIDED DIFFUSION

The existence of a drift field can result in motion of charged impurities in the direction of the electric field. Assuming that the random scattering of impurities by the lattice gives a resultant drift velocity v, the equation of motion due to an electric field E is given by

$$F = qE = m^* \, dv/dt + \alpha \, v \qquad \text{... (6.23)}$$

where F = force on impurity ion, m^* = effective mass and α = proportionality factor. The direction of the force obviously depends on the sign of the charge on the impurity ion.

In the steady state when a steady drift velocity v_d is attained

$$v_d = q \, Z \, E/\alpha = \mu \, E \qquad \text{... (6.24)}$$

where qZ = charge of impurity ion, $\mu = q \, Z/\alpha$ = mobility of impurity ion. If the motion of impurity ions due to drift and diffusion are considered to be independent then the flux density in presence of drift can be written as

$$j = -D \, (\partial N/\partial x) + \mu NE \qquad \text{... (6.25)}$$

In the case of diffusion of a n-type (donor) impurity, the ionized donor produces electrons which can diffuse much faster than the donors thus giving rise to an electric field as shown in figure 6.15 which further aids the motion of positively charged donors. Thus the motion of donors is aided by the space charge created by the diffusing electrons giving an enhanced diffusion coefficient.

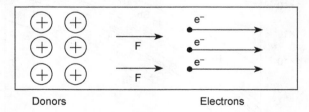

Fig. 6.15 Electric field–aided diffusion

The diffusion coefficient is related to the mobility by

$$D = (kT/q) \, \mu \qquad \text{... (6.26)}$$

Further in equilibrium the flow of electrons due to drift is balanced by the flow due to diffusion.

Thus
$$\mu n E = -D\,(\partial n/\partial x) = (kT/q)\mu\,(\partial n/\partial x) \qquad \ldots (6.27)$$
Simplifying
$$E = -(kT/q)(1/n)(dn/dx) \qquad \ldots (6.28)$$
Substituting in eqn. (6.25)
$$j = -D\,(1 + dn/dN)\,(\partial n/\partial x) \qquad \ldots (6.29)$$
Thus the impurities move with an effective diffusion coefficient Deff
$$D_{eff} = D\,(1 + dn/dN) \qquad \ldots (6.30)$$
For an n-type impurity
$$n/n_i = N/2\,n_i + [(N/2n_i)^2 + 1]^{1/2}$$
so that
$$dn/dN = 1/2\,[1 + \{1 + (N/2n_i)^2\}^{-1/2}]$$

Thus there can be a substantial increase in the effective diffusion coefficient with doping concentration, by even a factor of 2. This has been witnessed experimentally for substitutional diffusers.

6.8 DIFFUSION SYSTEMS

The choice of dopants in Si has been discussed. There are broadly 2 types of diffusion systems employed (i) open tube and (ii) closed tube. There are some general requirements for diffusion systems. These are:

(*a*) The surface concentration should be capable of being controlled over a wide range up to the solid solubility limit

(*b*) The diffusion process should not result in any damage to the surface

(*c*) The dopant remaining after diffusion should be capable of easily removed and

(*d*) The system should be reproducible and capable of handling a large number of wafers simultaneously.

(*e*) The temperature control should provide a central flat zone with $\pm 1/2°C$ variation in Temperature.

A diffusion furnace is a carefully designed apparatus proficient of upholding uniform temperature between 600–1200°C with a feedback controller. The diffusion tube made of high purity fused silica must be handled with great care, one tube and slice carrier being used for each type of dopant to prevent contamination. The length of the tubes varies from 10 cm – 150 cm or more for industrial furnaces. For large tubes the insertion of the carrier is done mechanically from one end, the other end being used for flow of gases and dopants. The temperature of the furnace is gradually ramped up from 600°C after insertion of the wafers with a programmed temperature controller ramping up the temperature at a linear rate of 3–10°C/min. This is to avoid thermal shock to the wafers as well as to the tube and components. In practice the diffusion tube is always kept above 600°C and never permissible to cool to room temperature to avoid devitrification. A gas source diffusion system is shown in figure 6.16(a).

If the temperature is ramped down at a rate $T = T_0 - Ct$ where $T_0 =$ initial temperature and C = constant, it can be shown that this is equivalent to the wafers being subject to an additional time kT_0/CE_0 at the initial diffusion temperature where E_0 = activation energy for diffusion.

B Diffusion

The most common p-type impurity is Boron because of its high solid solubility which is $6 \times 10^{20}/cm^3$ as given in Table 6.3. However due to the large misfit factor of B of 0.254 which introduces strain- induced defects, the actual upper limit is $5 \times 10^{19}/cm^3$. Diffusion systems for Boron in Si are summarized in Table 6.3.

Table 6.3 Diffusion system for Boron

Impurity source	Room Temperature	Temperature (°C)	Impurity Conc. range
BN	solid	950–1100°C	High & Low
BCL$_3$	gas	Room Temp.	High & Low
B$_2$H$_6$	gas	Room Temp.	High & Low

Solid, liquid and gaseous sources are available for B diffusion. One of the most common is Boric Oxide (B$_2$O$_3$). A preliminary reaction with B$_2$O$_3$ gives:

$$2 B_2O_3 (s) + 3 Si (s) \longrightarrow 4 B (s) + 3 SiO_2 (s)$$

The Si and B$_2$O$_3$ are kept at the same temperature and pre-deposition is carried out in N$_2$ ambient with 2–3% O$_2$. The temperature of the B$_2$O$_3$ controls the surface concentration of B as shown in (Fig. 6.16(c)). Excessive amounts of B$_2$O$_3$ leads to the formation of B skin which is difficult to remove. Slices are thus exposed to B$_2$O$_3$ source for a short time to form a glassy layer on the Si surface. The source is then removed and drive-in diffusion carried out in an oxidizing ambient.

This protects the surface against impurities. This process gives a 2 step-diffusion profile. BN slices slightly larger than the Si wafers can be used which can be sandwiched between Si slices with a spacing of 2–3 mm. These must be pre-oxidized at 750–1100°C to form a thin skin of B$_2$O$_3$ on the surface which forms the diffusion source:

$$4 BN (s) + 7 O_2 (g) \longrightarrow 2 B_2O_3 (s) + 4 NO_2 (g)$$

6.8 Diffusion Systems

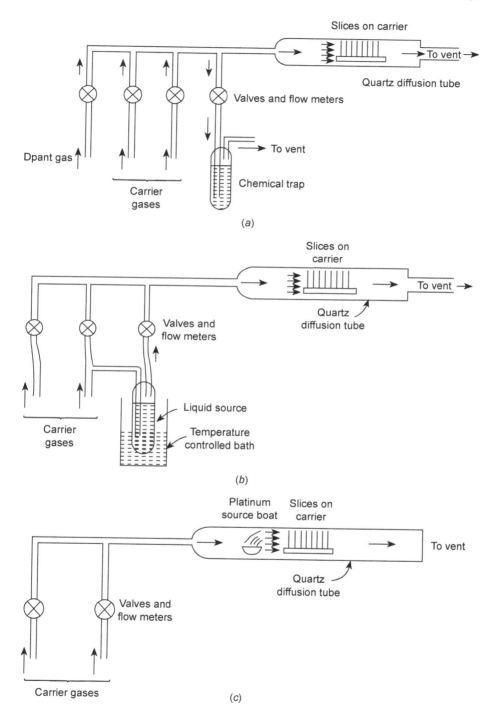

Fig. 6.16 (a) Gas source (b) liquid source (c) solid source diffusion system

No carrier gas is required but a flow of 1 l min of dry N_2 prevents back diffusion of contaminants. This process is extremely reproducible with excellent uniformity across the wafers. To avoid sticking, BN in a silica matrix is often used which also reduces B skin formation.

In thick film technology frequently used for the fabrication of solar cells, mixtures of B_2O_3 and SiO_2 in a polyvinyl alcohol solvent are used as spin-on sources. Mixtures of carborane and alkylsiloxane which have better viscosity control have also been used. An initial bake out is required before diffusion to convert the components into B_2O_3 and SiO_2.

Gaseous sources Fig. 6.16(a) which are used are diborane (B_2H_6) and BCl_3 which give the following reactions:

$$B_2H_6 (s) + 3\ O_2 (g) \xrightarrow{300°C} B_2O_3 (s) + 3\ H_2O\ (l)$$

$$4\ BCl_3 (g) + 3\ O_2 (g) \longrightarrow 2\ B_2O_3 (s) + 6\ Cl_2 (g)$$

P Diffusion

The activation energy is the same as for B but the misfit factor is small compared with B. High doping up to $5 \times 10^{20}/cm^3$ makes this an attractive system. The sources available are: Liquid sources: $POCl_3$, PCl_3 and PBr_3

$$4\ POCl_3\ (l) + 3\ O_2 (g) \longrightarrow 2\ P_2O_5 (s) + 6\ Cl_2 (g)$$

Table 6.4 Diffusion system for phosphorus in Si

Inpurity source	R.T. State	Temp. range (°C)	Impurity conc. range	Advantage	Disadvantage
$POCl_3$	Liquid	0–40	High & Low	Clean system; good control over wide range of impurity concentration	System geometry important
PCl_3	Liquid	170	High & Low	Can be used in non-oxidising diffusion	
PH_3	Gas	R.T.	High & Low	Accurate control by gas flow control	Highly toxie & explosive

Diffusion system for phosphorus in Si are summarised in table 6.4 of these the most popular is $POCl_3$. An oxidising gas mixture is used in the pre-deposition stage. The presence of O_2 reduces halogen pitting which becomes appreciable only for doping conc. $>10^{21}/cm^3$. Adjustment of bubbler temperature gives good control over surface concentration figure 6.16(b).

Gas Source: PH_3 with 99.9% O_2. The reaction is:

$$PH_3 (g) + 4\ O_2 (g) \longrightarrow P_2O_5 (s) + 3\ H_2O\ (l)$$

Sb Diffusion

This is used in special cases when the dopant impurity should be immobile under further processing because Sb has a relatively high diffusion activation energy of 3.95 eV. The sources available are:

Solid sources: Sb_2O_3 and Sb_2O_4 at 900°C

Liquid sources: Sb_3Cl_5 in a bubbler

In the last case Sb is transported as an oxide. Diffusion occurs through a glassy layer following surface reaction with Si.

As Diffusion

As has misfit factor = 0 with Si and hence does not give rise to strain on heavy doping. It is thus used for the fabrication of low resistivity epitaxial layers. It is highly toxic and hence the diffusion systems must be handled with extreme care. The sources used are:

Solid sources: $2\ As_2O_3\ (s) + 3\ Si\ (s) \longrightarrow 3\ SiO_2\ (s) + 4\ As\ (s)$

Gas sources: $2\ AsH_3\ (g) + 3\ O_2\ (g) \longrightarrow As_2O_3\ (s) + 3\ H_2O\ (l)$

Au Diffusion

Au is a very rapid diffuser in Si, almost 10^5 faster than B or P. It is used as a deep level recombination centre to reduce the minority carrier life-time and hence switching time in diodes and transistors. Prior to diffusion it is vacuum deposited on Si as a ~ 10 nm thick layer on the back surface of the wafer. Au-Si alloy forms resulting in damage to the Si surface. The diffusion time is typically 10–15 min at 800–1050°C and results in Au diffusion throughout the wafer. Au diffusion must be followed by rapid withdrawal and cooling to room temperature to prevent out-diffusion effects. Since gold doping is difficult to control it is being replaced by alternative techniques such as radiation-induced centers which can be area-selective with the dose and energy being easier to control.

6.9 OXIDE MASKING

Since semiconductor devices and ICs require selective area doping, masks are required to prevent diffusion in certain areas. The properties of SiO_2 are ideal for acting as a mask since the diffusion coefficients of most impurities such as B, P and As are orders of magnitude smaller in SiO_2 than in Si. However SiO_2 cannot act as a mask for Ga and Al, the latter attacking SiO_2 reducing it to Si. SiO_2 can be grown easily on Si by thermal oxidation and windows etched in it by photolithography such that the remaining areas act as masks. The windows permit impurity diffusion to form p-n junctions as required. The minimum thickness of the SiO$_2$ layers to act as a mask for a particular diffusion process must be determined. The diffusion process in SiO$_2$ can be considered to consist of two steps: in the first the dopant impurities react with the SiO$_2$ to form a glass. As the process continues the glass thickness increases until it penetrates the entire thickness of the oxide. At this point the second step commences – the impurity after diffusing through the glass reaches the glass – Si interface and starts diffusing into the Si. The first step is when the SiO$_2$ is effective as a mask against a given impurity. The required oxide thickness depends on the diffusivity of the impurity in SiO$_2$. Typical diffusivities at 900°C, 1100°C and 1200°C are given in Table 6.5.

Table 6.5: Diffusivity at different temperature

Element	D at 900°C / cm^2/s	D at 1000°C / cm^2/s	D at 1100°C / cm^2/s
B	3×10^{-19}	3×10^{-17} – 2×10^{-14}	2×10^{-16} – 5×10^{-14}
P	1×10^{-18}	2.9×10^{-16} – 2×10^{-13}	2×10^{-15} – 7.6×10^{-13}

Figure 6.17 shows the minimum thickness of dry-oxygen grown SiO$_2$ required as a mask against B and pass a function of temperature and time. It is noted

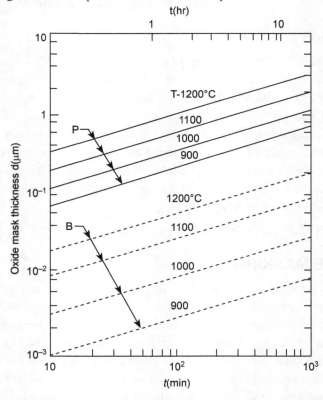

Fig. 6.17 Minimum thickness of SiO$_2$ required to mask against B and P diffusion

that P requires thicker masks for the same diffusing conditions since it has a higher diffusivity in SiO_2. For a given temperature the thickness of d varies as 1/2 since the diffusion length varies as $(Dt)^{1/2}$. An oxide mask thickness of 0.5 – 0.6 µm is adequate for most conventional diffusion steps.

6.10 IMPURITY REDISTRIBUTION DURING OXIDE GROWTH

During thermal oxidation dopant impurities are redistributed between the oxide and Si. This is because when two solid surfaces are in contact an impurity will redistribute between the two until it reaches equilibrium. This depends on several factors including the segregation coefficient k which is defined as in the case of zone melting in **Chapter 3**, as k = equilibrium concentration of impurity in Si/equilibrium concentration of impurity in SiO_2.

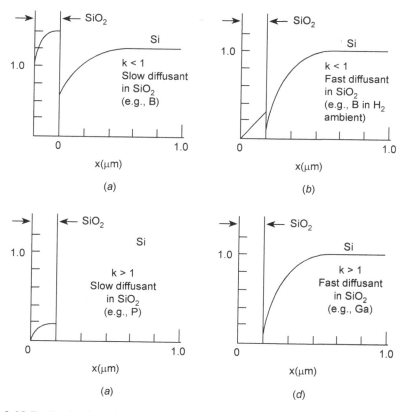

Fig. 6.18 Redistribution of Impurity between thermal oxide and Si (a) and (b) K < 1; (c) and (d) k < 1

Another factor is the rapid diffusion of the impurity through the oxide and escape into the ambient. This will depend on the diffusivity of the impurity in the oxide. A third factor is the growth of the oxide into the Si and the consequent motion of the Si-oxide interface. Thus the redistribution will depend on the rate of movement of the oxide in comparison with the rate of diffusion of the impurity through the oxide. Since the oxide layer is about twice as thick as the Si it replaces the same impurity will be redistributed in a larger volume thus resulting in depletion of the impurity from Si even if k = 1.

Four distinct cases may arise:

1. k < 1 : The oxide takes up the impurity which diffuses slowly through the oxide. e.g. B with k = 0.3. Consequently there is build-up of impurity in the oxide (Fig. 6.18(a))
2. k < 1 : The oxide takes up the impurity which diffuses rapidly out through the oxide. e.g. B heated in H ambient, as H in SiO_2 enhances the diffusivity of B (Fig. 6.18(b))
3. k > 1 : The oxide rejects the impurity and the diffusivity of the impurity in SiO_2 is slow resulting in build-up at the Si interface e.g. k = 10 for P, Sb and As (Fig. 6.18(c))
4. k > 1 : The oxide rejects the impurity and the diffusivity of the impurity in SiO_2 is rapid so that the impurity escapes from the solid into the gaseous ambient that there is overall a depletion of the impurity e.g. Ga with k = 20 and a fast diffuser in SiO_2 (Fig. 6.18(d))

In practice redistribution effects are important for B with the surface concentration being reduced to 50% of its value in the absence of redistribution. For P the overall effect is negligible since the redistribution and diffusion effects cancel each other out. The impurities in the oxide are hardly electrically active but they affect processing and device properties. The oxidation rate is affected by high dopant concentrations in Si. Non-uniform distribution of impurities in the oxide affects the interface- state properties.

6.11 LATERAL DIFFUSION

Diffusion of impurities into a semiconductor slice being treated as a 1-dimensional problem is valid since the horizontal dimensions are much larger than the vertical diffusion depth. This is true except at the edge of the oxide diffusion mask where the impurities can diffuse laterally below the oxide mask. It is found that the ratio of lateral to vertical diffusion is between 65–70%. This obviously limits the proximity between adjacent windows in the mask and poses one limit to device miniaturization.

A 2-dimensional diffusion equation is required to solve this problem. Numerical solutions of the problem for different initial and boundary conditions are shown in figure 6.19. Contours of constant doping concentration for a constant-surface-concentration diffusion N/N_S are shown assuming that the diffusion coefficient is independent of concentration. The contours give the location of junctions formed by diffusion into a wafer with various doping concentrations. The x and y axes are normalized with respect to $(Dt)^{1/2}$. Taking a value of $N/N_S = 10^{-4}$ the appropriate constant-concentration curve shows that the vertical penetration is 2.8 units compared with a horizontal penetration of 2.3 units that the vertical penetration is 2.8 units compared with a horizontal penetration of 2.3 units.

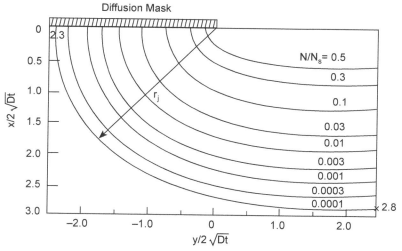

Fig. 6.19 Lateral diffusion effect at the edges of an oxide mask window

Lateral diffusion causes p-n junctions to have cylindrical edges through a radius of curvature r_j. If the mask has sharp corners the shape of the junctions near the corners will be roughly spherical. Cylindrical and spherical junction regions have higher curvature and hence higher avalanche breakdown voltages than for planar junctions with the same background doping concentrations.

6.12 DIFFUSION IN POLYSILICON

Polysilicon films are typically used in VLSI as a gate or as an intermediate conductor in two-level systems. Since the gate electrode is over a thin oxide (15 nm to 150 nm thick), it is imperative that dopant atoms in the polysilicon do not diffuse through the gate oxide. Polysilicon films are generally deposited at a low temperature without doping elements. After the gate region is defined, the polysilicon film is doped by diffusion (from a doped-oxide source or gas source) or by ion implantation.

Impurity diffusion in polysilicon film can be expounded qualitatively by a grain-boundary model. A polysilicon film is composed of single crystallites of changing sizes that are separated by grain boundaries. The diffusivity of impurity atoms that migrate along grain boundaries can be up to 100 times larger than that in a single crystal lattice. In addition, experimental results indicate that impurity atoms inside each crystallite have diffusivities either comparable to or a factor of 10 larger than those found in the single crystal. The diffusivity in a polysilicon film therefore depends strongly upon the structure (grain size, etc) and texture. These are in turn functions of the film deposition temperature, rate of deposition, thickness, and composition of the substrate. Therefore, it is difficult to predict diffusion profiles in Polysilicon. Diffusivities are characteristically estimated from junction depths and surface concentrations are determined experimentally.

6.13 MEASUREMENT TECHNIQUES

The results of a diffusion process can be evaluated by three parameters:

(1) Junction depth.

(2) Sheet resistance.

(3) Dopant profile.

Table 6.6 summarizes some of the more commonly used techniques:

Table 6.6 Comparison of various techniques

Technique	Sensitivity (atoms/cm^3)	Spot size	Quantity measured
Groove and stain	~10^{12}	~1 mm	Junction depth
Capacitance voltage plotting (C-V)	~10^{12}	~0.5 mm	Electrically active dopant levels versus depth
Four point probe	~10^{12}	~0.5 mm	Sheet resistance can be converted to average resistivity if the junction depth is known
Secondary ion mass spectroscopy (SIMS)	~10^{16}	~1 μm	Atomic concentration and type of contamination and depth profile
Spreading resistance probe (SRP)	~10^{12}	0.1 mm	Electrically active dopant versus depth

6.13.1 Staining

Junction depths are commonly measured on an angle-lapped (1° to 5°) sample chemically stained by a mixture of 100 c.c. HF (49%) and a few drops of HNO$_3$. If the sample is subjected to strong illumination for one to two minutes, the p-type region will be stained darker than the n-type region, as a result of a reflectivity difference of the two etched surfaces. The location of the stained

6.13 Measurement Techniques

junction depends on the p-type concentration level and sometimes on the concentration gradient. If the surface is then etched with a solution of HF+ HNO$_3$ (few drops in 100 ml) the p-type region is stained darker than the n-type region. In general, the stain boundary corresponds to a concentration level in the range of mid-10^{17} atoms/cm^3.

The junction depth is the position below the surface x_j where the dopant concentration matches the background concentration. It can be found by forming a groove on the semiconductor surface with a tool of radius R_0. The junction depth is then given by

$$x_j = (R_0^2 - a^2)^{1/2} - (R_0^2 - b^2)^{1/2} \quad \ldots (6.31)$$

where a and b are as indicated in Fig. 6.20. If $R_0 \gg a, b$, then

$$x_j \cong \frac{(a^2 - b^2)}{2R_0} \quad \ldots (6.32)$$

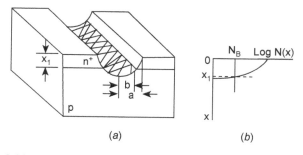

Fig. 6.20 Measurement of junction depth by groove and staining

6.13.2 Capacitance-Voltage Plotting (C-V)

The reverse bias capacitance of a p-n junction can be used to measure impurity levels. The technique requires that a shallow highly doped junction to be formed above the impurity profile to be measured. The highly doped junction must be of opposite conductivity type to the impurity being measured (n+ on p or p+ on n). The voltage on the junction is ramped and the instantaneous capacitance is measured. The capacitance can be converted to electrically active impurity concentration by

$$C(V) = \left[\frac{q\varepsilon_s C_B}{2}\right]^{1/2} \left[V_0 \mp \left(V_R - \frac{2kT}{q}\right)\right]^{1/2} \quad \ldots (6.33)$$

where, ε_S is the permittivity of silicon, C_B is the background concentration, V_0 is the junction potential, and V_R is the reverse bias on the junction.

The limitations of the technique are:

The diffusion being measured must be deeper than the shallow diffusion used for measurement plus the zero bias depletion width. More commonly this technique is applied on an MOS capacitor structure removing the shallow junction restrictions.

6.13.3 Four Point Probe (FPP)

The four-point-probe technique is used to measures the sheet resistance of a diffused junction. The FPP technique is illustrated in figure 6.21.

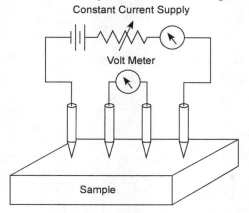

Fig. 6.21 four point probe for sheet resistance measurement.

In the system shown in figure 6.21 a constant current is supplied to the outer two probes and the voltage drop is measured across the inner two probes. The sheet resistance is given by

$$\rho = \frac{\pi}{\ln 2}\frac{V}{I}x_j = 4.532\frac{V}{I}x_j = R_S x_j \quad \ldots (6.34)$$

where, I is a forced current, V is the measured voltage drop and R_s is the sheet resistance.

For a diffused junction the concentration varies with depth so the average resistivity is given by

$$\rho = \frac{x_j}{q}\int_0^{x_j}\frac{1}{C(x)\mu(x)}dx \quad \ldots (6.35)$$

where, C(n) concentration of q is the electron and µ(x) is the mobility. Equation 6.34 can be numerically evaluated for diffused gaussian and error function profiles with varying background concentrations. The resulting curves of surface concentration versus average resistivity are known as Irvin curves after the first researcher to present the technique.

6.13.4 Secondary Ion Mass Spectroscopy (SIMS)

SIMS bombards the segment of the wafer to be investigated with a beam of energetic inert ions. The atomic composition is determined by mass analysis of ion beam sputters from the wafer surface.

Continuous bombing of a surface by the ion beam results in etching so that a depth profile can be determined. The minimum spot size for SIMS is relatively small at 1 μm. The drawbacks to SIMS are relatively low sensitivity – approximately 5×10^{16} atoms/cm^3, the equipment is expensive and difficult to use only the largest semiconductor companies can have the technique in-house, and SIMS is destructive. SIMS does however provide atomic concentration versus depth and allow atomic species to be identified. Because of SIMS unique abilities the technique is widely used to analyze doping profile and also to look for contamination.

6.13.5 Spreading Resistance Probe (SRP)

The wafer is first bevel ground and polished along a shallow angle to perform an SRP analysis. A tiny set of conductive needles are dragged along the beveled surface - see figure 6.22.

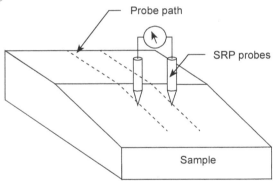

Fig. 6.22 Spreading resistance probe analysis

The voltage drop across the needles is measured by a known current applied between the needles. The resistivity in a small volume under the needle is given by

$$\rho = 2R_{SR} a \quad \ldots (6.36)$$

where, R_{SR} is the spreading resistance value and a is a geometric factor determined by measuring a sample of known resistivity.

The carrier concentration can be calculated by the relationship between concentration and resistivity. If the motion of the probes across the bevel is well controlled then the bevel angle is known so the profile versus depth can be calculated. SRP has a couple of limitations:

1. Only the electrically active dopant is considered so dopants must be activated prior to SRP (it is in contrast to SIMS that measures atomic concentration regardless to electronic state).
2. SRPs require a fairly large measurement area.
3. SRP is destructive and very technique sensitive.

6.14 SUMMARY

This chapter reviewed the physics of diffusion and presented Fick's laws, the relations that govern diffusion. Two particular solutions were presented corresponding to drive-in and pre-deposition diffusion. The atomistic mechanisms of diffusion were presented along with heavy doping effects. The details of diffusion for a variety of popular dopants were also discussed. At high doping concentrations the diffusion coefficient is no longer constant, but frequently depends on the local doping concentration and concentration gradient. A numerical tool, was introduced that allows the student to calculate dopant profiles in the presence of these nonlinear effects.

PROBLEMS

1. Assume that you have been asked to measure the diffusivity of a donor impurity in a new elemental semiconductor. What constants would you need to measure? What experiments would you attempt? Discuss the measurement techniques that you would use to measure the chemical and carrier profiles. What problems are likely to arise?
2. Assume that a wafer is uniformly doped. If a Schottky contact is formed on the surface, what would the C–V curve look like?
3. How sheet resistance of diffused junction is measured? Explain four point probe method for that.
4. Explain Fick law of diffusion in detail.
5. A diffusion furnace is ramped up (from 500°C) for 20 min, held at 100°C for 30 min, and ramped down to 500°C in 15 min. Calculate the effective diffusion time, assuming phosphorus in silicon.

REFERENCES

1. K.B. Kahen, "Mechanism for the Diffusion of Zinc in Gallium Arsenide," *Mater. Res. Soc. Symp. Proc.*, 163:681 (1990).
2. M.E. Greiner and J.F. Gibbons, "Diffusion of Silicon in Gallium Arsenide Using Rapid Thermal Processing: Experiment and Model," *Appl. Phys. Lett.* 44:740 (1984).

3. K.L. Kavanaugh, C.W. Magee, J. Sheets, and J.W. Mayer, "The Interdiffusion of Si, P, and in at Polysilicon Interfaces," *J. Appl. Phys.* 64:1845 (1988).
4. S. Yu, U.M. Gosele, and T.Y. Tan, "An Examination of the Mechanism of Silicon Diffusion in Gallium Arsenide," *Mater. Res. Soc. Symp. Proc.*, 163:671 (1990).
5. K.B. Kahen, D.J. Lawrence, D.L. Peterson, and G. Rajeswaren, "Diffusion of Ga Vacancies and Si in GaAs," in Mater. Res. Soc. Symp. Proc., 163:677 (1990).
6. J.J. Murray, M.D. Deal, E.L. Allen, D.A. Stevenson, and S. Nozaki, "Modeling Silicon Diffusion in GaAs Using Well-Defined Silicon Doped Molecular Beam Epitaxy Structures," *J. Electrochem. Soc.* 137(7):2037 (1992).
7 D. Sudandi and S. Matsumoto, "Effect of Melt Stoichiometry on Carrier Concentration Profiles of Silicon Diffusion in Undoped LEC Sl-GaAs," *J. Electrochem. Soc.* 136:1165 (1989).
8. L.B. Valdes, "Resistivity Measurements on Germanium for Transistors," *Proc. IRE* 42:420 (1954).
9. M. Yamashita and M. Agu, "Geometrical Correction Factor of Semiconductor Resistivity Measurement by Four Point Probe Method," *Jpn. J. Appl. Phys.* 23:1499 (1984).
10. D.K. Schroder, "Semiconductor Material and Device Characterization," Wiley-Interscience, New York, (1990).
11. L.J. Van der Pauw, "A Method for Measuring the Specific Resistivity and Hall Effect of Discs of Arbitrary Shape," *Phillips Res. Rep.* 13:1 (1958).
12. D.S. Perloff, "Four-point Probe Correction Factors for Use in Measuring Large Diameter Doped Semiconductor Wafers," *J. Electrochem. Soc.* 123:1745 (1976).
13. A. Diebold, M. R. Kump, J. J. Kopanski, and D. G. Seiler, "Characterization of two-dimensional dopant profiles: Status and review," *J. Vacuum Sci. Technol. B* 14:196 (1996).
14. J.S. McMurray, J. Kim, and C.C. Williams, "Direct Comparison of TwoDimensional Dopant Profiles by Scanning Capacitance Microscopy with TSUPRE4 Process Simulation," *J. Vacuum Sci. Technol. B.* 16:344 (1998).
15. M. Pawlik, "Spreading Resistance: A Comparison of Sampling Volume Correction Factors in High Resolution Quantitative Spreading Resistance," STP 960, American Society for Testing and Materials, Philadelphia, (1987).
16. R.G. Mazur and G.A. Gruber, "Dopant Profiles in Thin Layer Silicon Structures with the Spreading Resistance Profiling Technique," *Solid State Technol.* 24:64 (1981).
17. P. Blood, "Capacitance-Voltage Profiling and the Characterization of III-V Semiconductors Using Electrolyte Barriers," *Semicond. Sci. Technol.*, 1:7 (1986).

18. M. Ghezzo and D.M. Brown, "Diffusivity Summary of B, Ga, P, As, and Sb in SiO_2," *J. Electrochem. Soc.*, 120:146 (1973).

19. Z. Zhou and D.K. Schroder, "Boron Penetration in Dual Gate Technology," *Semicond. Int.* 21:6 (1998).

20. K.A. Ellis and R.A. Buhrman, "Boron Diffusion in Silicon Oxides and Oxynitrides," *J. Electrochem. Soc.* 145:2068 (1998).

21. T. Aoyama, H. Arimoto, and K. Horiuchi, "Boron Diffusion in SiO2 Involving High Concentration Effects," *Jpn. J. Appl. Phys.* 40:2685 (2001).

22. S. Sze, "VLSI Technology," McGraw-Hill, New York, (1988).

23. M. Uematsu, "Unified Simulation of Diffusion in Silicon and Silicon Dioxide," Defect Diffusion Forum, 38:237 (2005).

24. T. Aoyama, H. Tashiro, and K. Suzuki, "Diffusion of Boron, Phosphorus, Arsenic, and Antimony in Thermally Grown Silicon Dioxide," *J. Electrochem. Soc.* 146(5):1879 (1999).

25. M. Susa, K. Kawagishi, N. Tanaka, and K. Nagata, "Diffusion Mechanism of Phosphorus fromPhosphorus Vapor in Amorphous Silicon Dioxide Film Prepared by Thermal Oxidation," *J. Electrochem. Soc.* 144(7):2552 (1997).

26. T. Aoyama, K. Suzuki, H. Tashiro, Y. Toda, T. Yamazaki, K. Takasaki, and T. Ito, "Effect of Fluorine on Boron Diffusion in Thin Silicon Dioxides and Oxynitrides" *J. Appl. Phys.* 77:417 (1995).

27. T. Aoyama, K. Suzuki, H. Tashiro, Y. Tada, and K. Horiuchi, "Nitrogen Concentration Dependence on Boron Diffusion in Thin Silicon Oxynitrides Used for Metal-Oxide-Semiconductor Devices," *J. Electrochem. Soc.* 145:689 (1998).

28. K.A. Ellis and R.A. Buhrman, "Phosphorus Diffusion in Silicon Oxide and Oxynitride Gate Dielectrics," *Electrochem. Solid State Lett.* 2(10):516 (1999).

29. F.J. Morin and J.P. Maita, "Electrical Properties of Silicon Containing Arsenic and Boron," *Phys. Rev.* 96:28 (1954).

30. W.R. Runyan and K.E. Bean, "Semiconductor Integrated Circuit Processing Technology," Addison-Wesley, Reading, MA, (1990).

31. P.M. Fahey, P.B. Griffin, and J.D. Plummer, "Point Defects and Dopant Diffusion in Silicon," *Rev. Mod. Phys.* 61:289 (1989).

32. T.Y. Yan and U. Gosele, "Oxidation-Enhanced or Retarded Diffusion and the Growth or Shrinkage of Oxidation-Induced Stacking Faults in Silicon," *Appl. Phys. Lett.* 40:616 (1982).

33. S. Mizuo and H. Higuchi, "Retardation of Sb Diffusion in Si During Thermal Oxidation," *J. Appl. Phys. Jpn.* 20:739 (1981).

34. A.M.R. Lin, D.A. Antoniadis, and R.W. Dutton, "The Oxidation Rate Dependence of Oxidation-Enhanced Diffusion of Boron and Phosphorus in Silicon," *J. Electrochem. Soc.* 128:1131 (1981).

References

35. D.J. Fisher, "Diffusion in Silicon—A Seven-year Retrospective," *Defect Diffusion Forum* 241:1 (2005).

36. D.J. Fisher, "Diffusion in Ga-As and other III-V Semiconductors," *Defect Diffusion Forum* 157-159:223 (1998).

37. R.B. Fair, "Concentration Profiles of Diffused Dopants in Silicon," Impurity Doping Processes in Silicon, North-Holland, New York, (1981).

38. H. Ryssel, K. Muller, K. Harberger, R. Henkelmann, and F. Jahael, "High Concentration Effects of Ion Implanted Boron in Silicon," *J. Appl. Phys.* 22:35 (1980).

39. R. Duffy, V.C. Venezia, A. Heringa, B.J. Pawlak, M.J.P. Hopstaken, G.J. Maas, Y. Tamminga, T. Dao, F. Roozeboom, and L. Pelaz, "Boron Diffusion in Amorphous Silicon and the Role of Fluorine," *Appl. Phys. Lett.* 84 (21):4283 (2004).

40. A. Ural, P.B. Griffin, and J.D. Plummer, "Fractional Contributions of Microscopic Diffusion Mechanisms for Common Dopants and Self Diffusion in Silicon," *J. Appl. Phys.* 85(9):6440(1999).

41. J. Xie and S.P. Chen, "Diffusion and Clustering in Heavily Arsenic-Doped Silicon—Discrepancies and Explanation," *Phys. Rev. Lett.* 83(9): 1795 (1999).

42. R. B. Fair and J. C. C. Tsai, "A Quantitative Model for the Diffusion of Phosphorus in Silicon and the Emitter Dip Effect," *J. Electrochem. Soc.* 124:1107 (1978).

43. M. UeMatsu, "Simulation of Boron, Phosphorus, and Arsenic Diffusion in Silicon Based on an Integrated Diffusion Model, and the Anomalous Phosphorus Diffusion Mechanism," *J. Appl, Phys.* 82(5): 2228 (1997).

44. R.J. Field and S.K. Ghandhi, "An Open Tube Method for the Diffusion of Zinc in GaAs," *J. Electrochem. Soc.* 129:1567 (1982).

45. L.R. Weisberg and J. Blanc, "Diffusion with Interstitial-Substitutional Equilibrium. Zinc in Gallium Arsenide," *Phys. Rev.* 131:1548 (1963).

46. S. Reynolds, D.W. Vook, and J.F. Gibbons, "Open-Tube Zn Diffusion in GaAs Using Diethylzinc and Trimethylarsenic: Experiment and Model," *J. Appl. Phys.* 63:1052 (1988).

7

Ion Implantation

7.1 INTRODUCTION

Ion implantation is an alternative to diffusion process which changes the physical and electronic properties of a material by forcibly embedding different types of ions into the material. The technique dates back to the 1940's when it was developed at Oak Ridge National Laboratory. Since then the technique has applied in several materials processing processes. In the 1970's the use of ion implantation to modify the electrical properties of semiconductors, metals, insulators and ceramics became extremely popular. In diffusion process the transistor dimensions is not accurately controlled as the quantity of doping and the depth of doping is not quite controllable. The constant miniaturization of device dimensions in integrated circuit needs precised control on transistor dimensions. When the transistor dimensions were 5 micron (i.e. 5,000 nm) or so, if the variability in doping level is 100 nm, it was acceptable because it was only 2% but this much variability is not acceptable when the transistor itself is 100 nm of size. Ion implantation is a very prevalent process for VLSI because it arrange for more precise control of dopants (as compared to diffusion). With the reduction of device sizes to the submicron range, the electrical activation of ion-implanted species relies on a rapid thermal annealing technique, resulting in as little movement of impurity atoms as possible. Thus, diffusion process has become less significant than methods for introducing impurity atoms into silicon for making very shallow junctions, an important feature of VLSI circuits.

The main benefits over the diffusion are low temperature, more precise control and reproducibility of impurity doping, and shallow implant. However, owing to high-energy bombardment causing damaged crystal lattice, Rapid thermal annealing (RTA) at 400°C to 500°C is required to allow the implanted atom to stay at the right substitutional site, to repair crystal damage, and drive-in the implanted atom. Ion implantation permits introduction of the dopant in silicon that is controllable, reproducible and free from undesirable side effects. Over the past few years, ion implantation has been advanced into a very powerful tool for IC fabrication. Its attributes of controllability and reproducibility make it a very versatile tool, able to follow the trends to finer-scale devices. Ion implantation continues to find new applications in VLS technologies.

Characteristics of Ion Implantation

Implantation is done at relatively low temperatures, this means that doped layers can be implanted without disturbing previously diffused regions. This means a lesser tendency for lateral spreading

- Ion implantation provides much more precise control over the density of dopants deposited into the wafer, and hence the sheet resistance. This is possible because both the accelerating voltage and the ion beam current are electrically controlled outside of the apparatus in which the implants occur.
- The beam current can be measured accurately during implantation; a precise quantity of impurity can be introduced. Tins control over doping level, along with the uniformity of the implant over the wafer surface, make ion implantation attractive for the IC fabrication, since this causes significant improvement in the quality of an IC.
- The depth of penetration of any particular type of ion will increase with increasing accelerating voltage. The penetration depth will generally be in the range of 0.1 to 1.0 µm.
- After the ions have been implanted they are lodged principally in interstitial positions in the silicon crystal structure, and the surface region into which the implantation has taken place will be heavily damaged by the impact of the high-energy ions. The disarray of silicon atoms in the surface region is often to the extent that this region is no longer crystalline in structure, but rather amorphous. To restore this surface region back to a well-ordered crystalline state and to allow the implanted ions to go into substitutional sites in the crystal structure, the wafer must be subjected to an annealing process. The annealing process usually involves the heating of the wafers to some elevated temperature often in the range of 1000°C for a suitable length of time such as 30 minutes. Laser beam and electron-beam annealing are also employed. In such annealing techniques only the surface region of the wafer is heated and re-crystallized. An ion implantation process is

- often followed by a conventional-type drive-in diffusion, in which case the annealing process will occur as part of the drive-in diffusion.
- Ion implantation is a substantially more exclusive process than conventional deposition diffusion, both in terms of the cost of the equipment and the throughput.
- Due to precise control over doping concentration, it is possible to have very low values of dosage so that very large values of sheet resistance can be obtained. These high sheet resistance values are useful for obtaining large-value resistors for ICs. Very low-dosage, low-energy implantations are also used for the adjustment of the threshold voltage of MOSFET's and other applications.
- Now-a-day's large diameter wafers are feasible. Large size wafers are necessary for VLSI. This makes the task of uniformly implanting a wafer increasingly difficult. This in turn has effect on sheet resistance. Ion implantation is basically clean process because contaminant ions are separated from the beam before they hit the target. There are still several sources of contamination possible near the end of the beam line, which can result in contaminant dose up to 10 percent of the intended ion dose, for example, metal atoms knocked from chamber walls, water holder, masking aperture and so on.
- Annealing is required to repair lattice damage and put dopant atoms on substitutional site where they will be electrically active. The success of annealing is often measured in terms of the fraction of the dopant that is electrically active, as found experimentally using a Hall Effect technique. For VLSI, the challenge in annealing is not simply to repair damage and activate dopant, but to do so while minimizing diffusion so that shallow implants remain shallow.

7.2 ION IMPLANTER

Common ion implanter has much the same as the linear accelerator. Indeed the early history of ion implanter made much progress when research in nuclear physics shifted to high energy that laboratory machine was no longer useful and was instead used to investigate the relatively low energy of ion-solid interaction that is useful for ion implantation. A schematic of an ion implanter is shown in figure 7.1. Basically, the ion implant system consists of several systems, which are gas system, electrical system, vacuum system, control system, and beam line system. These systems will be discussed in details in the following sub-sections.

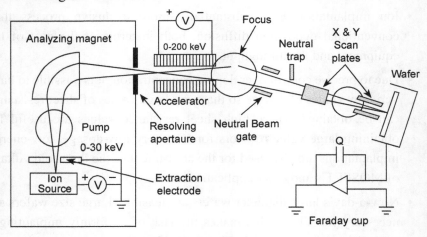

Fig. 7.1 Schematic of an ion implanter

7.2.1 Gas System

The basic requirement of an ion implanter is the source of ion of adequately high energy. Either a solid source is vaporized or gas source is conventionally used to delivery material for the ion implanter. Arsine, phosphine, diborane, and boron trifluoride (BF_3) are gas sources and phosphorus pentoxide (P_2O_5) is a solid. The common gas sources are extremely toxic and have been used in dilute mixture 15% in hydrogen gas in high pressure more than 400 psi cylinder. Owing to safety concern, solid sources of elemental boron, arsenic or phosphorus are at time preferred. The main benefit of solid is that it can be evaporated and implanted. New gas such as zeolite matrix which acts as a molecular sieve to absorb and store gas in cylinder below atmospheric pressure reduces the risk of release and explosion.

7.2.2 Electrical System

High voltage and current electrical system is required for ion implanter. High voltage DC power is needed to accelerate ion. Up to 200 kV DC power supply system is equipped in an implanter. The ion in the ion source is generated either a hot filament or RF plasma system. The hot filament requires large current and a few thousand volt bias power supply, whereas the RF ion source system needs a thousand watt of RF power. The analyzer magnet needs a high current to generate the magnetic field strong enough to deflect ion trajectory and helps to select right ions and created ultra-pure ion beam.

7.2.3 Vacuum System

The beam line must be in high vacuum condition to minimize collision between energetic ions with neutral gas molecules along the ion trajectory. Collision can cause ion scattering loss and creation of unwanted ion species for ion implantation. The vacuum requirement is 10^{-7} torr for beam line system. The high vacuum requirement can be achieved by combining the cryo-pump, turbo pump, and dry pump.

The dangerous gases are used in ion implantation. Thus, the exhaust of the vacuum system of ion implanter must be separated from the exhaust of other systems. The exhaust gas needs to go through a burn box and a scrubber before it can be released into the atmosphere. In the burn box, the combustible and explosive gases are neutralized with oxygen in high temperature flame. In the scrubber, flushing water dissolves corrosive gases and burn dust passing it.

7.2.4 Control System

The ion implanter needs to precisely control the ion beam energy, current, and ion species. The implanter needs to control the mechanical parts such the robot for wafer loading and unloading, and control wafer movement in order to achieve uniform implantation across the wafer. The throttle valves are controlled according to the pressure setting point to maintain system pressure.

7.2.5 Beam Line System

The ion beam line system is the most important part of an ion implanter. It consists of an ion source, extraction electrodes, mass analyzer, post acceleration, plasma flooding system, and end analyzer. These components are discussed as such.

7.2.5.1 Ion Source

Dopant ion is generated from ion source through ionization discharge of the atom or molecule of dopant vapor or gaseous dopant chemical compound. The hot filament ion source is one of the most commonly used ion source. The thermal electron hot filament is accelerated by the arc power supply to attain energy high enough to dissociate and ionize dopant gas molecule or dopant atom. Magnetic field in the ion source forces the electron into gyro-motion, which helps electron to travel longer distance and increase the probability of its collision with dopant molecule to generate more dopant ions. Other types of ion source are RF ion source and microwave ion source. The RF ion source uses inductive coupling of the RF power to ionize the dopant ions. The microwave ion source uses electron cyclotron resonance to generate plasma and ionized dopant ions.

7.2.5.2 Extraction

An extraction electrode with negative bias draws the positive ion out from the plasma in the ion source and accelerates it to adequately high 50 keV energy. It is a requirement for ion to attain high energy before the analyzer magnetic field can select the right type of ion species. When the dopant ions accelerate toward the extraction electrode, some of the ions pass through the slit and continue to travel along the beam line. Dome hit the extraction electrode surface, which generates X-ray and excites some secondary electrons. A suppression electrode with sufficiently lower electrical potential up to 10 kV than the extraction electrode is used to prevent these electrons being accelerated back to the ion source that would cause damage. All electrodes are shaped with a narrow slit through which ions are extracted as a collimated ion flux forming an ion beam.

7.2.5.3 Mass Analyzer

In magnetic field, the charge ion starts to rotate from the magnetic force, which is always perpendicular to the direction of the charged ion. For the fixed magnetic field strength and ion energy; the gyro radius is related only to the mass to charge ratio or m/q of the charge ion. This property had been used for isotope separation to get enriched uranium^{235} for making nuclear bomb. In most of the ion implanter, mass analyzer is used to select precisely the right of ion for implantation and weed out unwanted ion species. Species with less m/q value will defect more and will not pass through slit of the mass analyzer. Likewise, species with large m/q ratio will be stop too. A mass analyzer of an ion implanter is shown in figure 7.2.

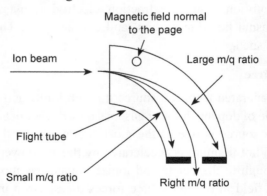

Fig. 7.2 Mass analyzer of Ion Implanter

Boron trifluoride (BF_3) is one of the frequently used materials for p-type implant. In the plasma, there are combinations of dissociative, ionized radicals, and recombined molecules. Boron has two isotopes, which are boron-10 ^{10}B (19.9%) and boron-11 ^{11}B (80.1%). Thus, in the plasma state are a number of ion types, which are ^{10}B (10g), ^{11}B (11g), ^{10}BF (29g), ^{11}BF (30g), F_2 (38g), $^{10}BF_2$ (48g), and $^{11}BF_2$ (49g). The figure in the parenthesis indicates the atom weight or molecular weight. For p-well implantation, the lighter weight $^{10}B^+$ is preferred because it can penetrate deeper into silicon substrate. For shallow junction implant $^{11}BF_2^+$ is preferred because of it large size and heavy weight. At the lowest energy level, an ion implanter can provide $^{11}BF_2$ ion for shallowest p-type junction implant.

7.2.5.4 Post Acceleration

After the analyzer chooses the correct ion species, the ion goes through the post acceleration section, where the beam current and final ion energy are controlled. The ion beam current is controlled by a pair of adjustable vanes, and the ion energy by post acceleration electrode potential, Ion beam focus and beam sharp are controlled in this part by defining apertures and electrodes.

For the high energy ion implants, which are mainly in the well and buried layer, it requires several high voltage acceleration electrodes connected in series along the beam line in order to accelerate the ions to several MeV. For ultra-shallow ion implant like the p-type boron implant, the electrode of post acceleration is connected in reversed way so that ion beam is decelerated instead accelerated when passing the electrode. It can generate a pure ion beam with energy as low as 500 eV.

To avoid asymmetrical distribution of implant for all devices, the wafer often rotates during implant or is implanted in four separate rotations. To completely avoid shadowing effect requires that the implant to be done at zero tilt angle. Alternatively, the x-y scan plates can direct the beam to scan the surface of wafer for ion implantation.

Ion implantation provides a very precise way to introduce a specific dose or number of dopant atoms into the silicon lattice. This is because the electrical charge on the ion allows it to be counted by Faraday cup. In spite of the preciseness in which the dose can be controlled.

7.3 ION IMPLANT STOP MECHANISM

When ion bombards and penetrates the silicon substrate, it collides with lattice atom. The ion gradually loses its energy and eventually stops inside the crystal lattice. There are two stop mechanisms namely nuclear stopping and electronic stopping. When ion collides with nuclei of the lattice atom it can be scattered significantly and transfer its energy to the atom in lattice. This type of stopping is called nuclear stopping. In the hard collision, lattice atom can get enough energy to break free from the lattice binding energy, which causes lattice disorder and damage crystal structure. If the ion collides with electron of the atom this type of collision, which is a soft collision, will not change the path of the ion and the energy of the ion significantly. It will not cause crystal damage and the range of penetration will be long. This type of collision is called electronic stopping.

The type of collision of ion in silicon lattice can be random collision, channel collision, and back scattering. Random collision and back scattering would be the nuclear stopping, while channel collision will have electronic stopping. Figure 7.3 illustrates the type of stopping mechanism. When projected ion hits the nucleus, it constituents nuclear stopping. When the projected ion entersinto substrate, it gets aligned with the gap between the host atoms, and they travel a large distance before finally coming to rest. This phenomenon is called ion channeling.

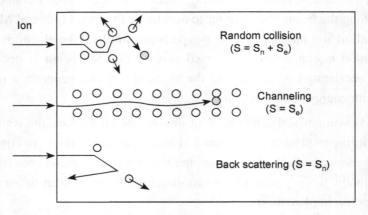

Fig. 7.3 Stopping Mechanism

The total stopping power S of the target, which is defined the energy loss E per unit path length of the ion x, would be consisted of two components namely the nuclear stop and electronic stop. Figure 7.4 shows the relationship between stopping power and velocity of ion for different types of stopping. Mathematically, it can be expressed as

$$S = \left(\frac{dE}{dx}\right)_{nuclear} + \left(\frac{dE}{dx}\right)_{electronic} \quad \ldots (7.1)$$

7.3 Ion Implant Stop Mechanism

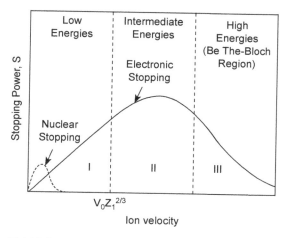

Fig. 7.4 Relationship between stopping power and velocity

Nuclear stopping is caused by a collision between two atoms, and can be described by classical kinematics. If the atoms were bare nuclei, then at a separation r, the columbic potential $V_c(r)$ between them would be

$$V_c(r) = \frac{q^2 Z_1 Z_2}{4\pi\varepsilon_0 r} \quad \ldots (7.2)$$

Where Z_1 and Z_2 are the atomic numbers of the atoms in reality, electrons screen the nuclear charge and a screening function must be included. With the presence of electron spinning around the nucleus, the screening function $f_s(r)$ have to be factor in. The screening functions (r) is defined as

$$f_s(r) = a_1 e^{-\frac{r}{b_1}} + a_2 e^{-\frac{r}{b_2}} + a_3 e^{-\frac{r}{b_3}} + \ldots \quad \ldots (7.3)$$

Thus, the coulombic potential V(r) is then equal to

$$V(r) = V_c(r) f_s(r) \quad \ldots (7.4)$$

With this interaction potential, the equation of motion of the atom can be integrated to yield the scattering angle for any incident ion trajectory, although it must be done numerically for realistic potentials With the center of mass frame, it simplifies the derivation of equation involved. Figure 7.5 shows the view of ion scattering showing its relationship between impact parameter and scattering cross section. Based on the Center of Mass Frame (CMF) a function of the impact parameter (p), the energy loss [T(p)] by the ion related to scattered angle θ is given by

$$T(p) = \frac{4 M_1 M_2}{(M_1 + M_2)^2} E \sin^2\left\{\frac{\theta(p)}{2}\right\} \quad \ldots (7.5)$$

where M_1 and M_2 are the atomic mass numbers of ion and target atom respectively. The probability of having an impact parameter between p and p+dP is 2π pdp, which is also known as the differential scattering cross section $d\sigma$.

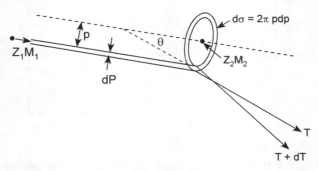

Fig. 7.5 Relationship between impact parameter and scattering cross section

Note# that based on center of mass frame, the velocity v_{imp} before impact is equal to the sum of the velocity after impact and the impacted atom, which $v_{imp} = v_1 + v_2$ where v_1 and v_2 are respectively equal to the after impact velocity of implant ion and velocity of atom receiving the impact. According to the center of mass, the after impact momentum of the implant ion is equal to the momentum of atom that receiving the impact.

The rate of energy loss to nuclear collision per unit path length is equal to the sum of energy loss for each possible impact parameter multiplied by the probability of that occurring collision. If the maximum possible energy transfer in a collision is T_{max} and there are N target atmosper unit volume then the nuclear stopping energy is

$$S_{nuclear} = \left(\frac{dE}{dx}\right)_{nuclear} = N \int_0^{T_{max}} T d\sigma \qquad \ldots (7.6)$$

where $d\sigma$ is the differential cross section. Nuclear stopping is elastic, and so energy lost by the incoming ion is transferred to the target atom that is subsequently recoiled away from its lattice site, thus creating a damage or defect site.

Electronic stopping is caused by the interaction between the incoming ion and the electrons in the target. The theoretical model is quite complex, but in the low energy regime, the stopping is similar to a viscous drag force and is proportional to the ion velocity. Electronic stopping is inelastic. The energy loss by incident ions is dissipated through the electron cloud into thermal vibrations of the target.

7.4 RANGE AND STRAGGLE OF ION IMPLANT

Ion implantation is a random process due to each ion follows its own random trajectory, scattering off the lattice silicon atom before its energy and coming to rest at some location as illustrated in figure 7.6. The reason ion implantation can be used successfully is because large numbers of ions are implanted so an average depth for the implanted dopants can be calculated.

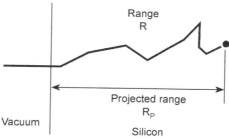

Fig. 7.6 Different ranges of ion implantation incident normal to the surface of silicon

The distributions of ion implanted in silicon for various types of dopant are shown in figure 7.7. Heavy ion such as antimony does not travel as far as the lighter ion like boron. The distribution of heavy ion is narrower than the distribution of lighter ion. The peak of the concentration of the ion is not near the surface of silicon. It is situated at an average distance away from the surface of silicon, which called the average projected range R_n.

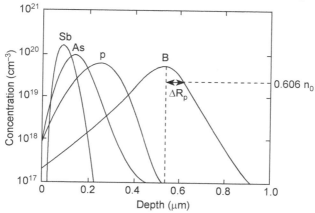

Fig. 7.7 Ion implanted distributions for various types of dopants in silicon lattice at energy of 200 keV.

Each implanted ion traverses a random path as it penetrates the target, losing energy by nuclear and electronic stopping. Since implantation doses are usually higher than 10^2 ions/cm², ion trajectories can be predicted employing statistical means. The average total path length is called the range (R), which is composed of both lateral and vertical motions as shown in figure 7.6. The average depth of the implanted ions is called the projected range (R_p) and the distribution of the implanted ions about that depth can be approximated as gaussian with a standard deviation R_p (or $\Delta\sigma_p$) as shown in figure 7.7. The lateral motion of the ions leads to a lateral gaussian distribution with a standard deviation $\Delta R_\perp$ (or $\Delta\sigma_\perp$). Far from the mask edge, the lateral motion can be ignored and $n(x)$, the ion concentration at depth x, can be written as:

$$n(x) = n_0 \exp\left\{-\frac{(x-R_p)^2}{2\Delta R_p^2}\right\} \quad \ldots (7.7)$$

where n_o is the peak concentration, R_p is the projected range, and ΔR_p is the standard deviation. If the total implanted dose is Q_T, integrating equation 7.7 gives an expression for the peak concentration n_o

$$n_0 = \frac{Q_T}{\sqrt{2\pi}\Delta R_p} \cong \frac{0.4 Q_T}{\Delta R_p} \quad \ldots (7.8)$$

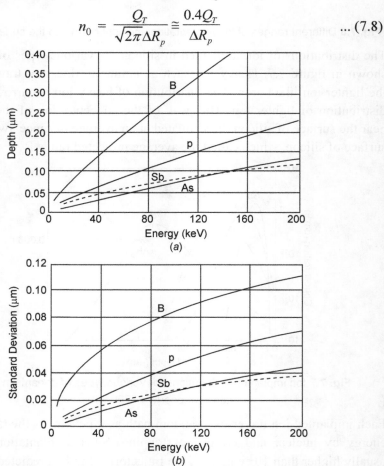

Fig. 7.8 (a) Plot of average range (b) standard deviation of various dopants in Silicon

7.4 Range and Straggle of Ion Implant

In general, an arbitrary distribution can be characterized in terms of its moments. The normalized first moment of an ion distribution is the projected range (R_p). The second moment is the standard deviation (ΔR_p). The third moment is the skewness (Γ) whereas the fourth moment, kurtosis, is designated β. Qualitatively skewness is a measure of the asymmetry of the distribution. Positive skewness places the peak of the distribution closer to the surface than R_p. Kurtosis is an indication of how flat the top of a distribution is a true gaussian distribution has a skewness of 0 and a kurtosis of 3.

Several different distributions have been employed to give a more accurate fit to the moments of an ion implant distribution than is possible using a gaussian. Figure 7.8 shows the plot and standard deviation for common dopants in silicon. The results show that the implant depth and standard deviation are linear for high energy implant. It is obvious that a heavy ion such as antimony and arsenic have let penetrating power shown by the depth of implant.

The total number of ion implanted is defined as dose and it simply by equation

$$Q_T = \int_{-\infty}^{\infty} n(x)d(x) = \int_{-\infty}^{\infty} n_0 \exp\left(-\frac{(x-R_p)^2}{2\Delta R_p^2}\right) dx \ldots (7.9)$$

The solution of the integration is equal to

$$Q_T = \sqrt{2\pi}\Delta R_p C_p \qquad \ldots (7.10)$$

Experimentally, it is relatively easy to disclose vertical atomic profiles. However, it is much more difficult to accurately assess lateral atomic profiles, the two dimensional projection near the window edge is of interest because it designates how many ions scatter under the window due to lateral straggle. Owing to the difficulty to experimentally measure the lateral dopant profile, a two-dimensional distribution is often assumed to be composed of the product of the vertical and lateral distribution.

$$n(x, y) = \frac{n_{vert}(x)}{\sqrt{2\pi}\Delta R_\perp} \qquad \ldots (7.11)$$

This equation describes the result of implanting at a single point on the surface to obtain the result of implanting through a mask window; Equation 7.11 can be integrated over the open areas where the ion beam can enter. Figure 7.9 displays the results for a 70 keV boron implant through 1μm slit in a thick mask showing that ions scatter well outside the open area. To minimize lateral scattering, masking layers are often tapered at the edge rather than perfectly abrupt, so that ions are gradually prevented from entering the silicon.

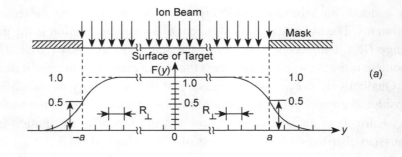

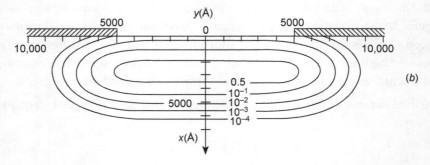

Fig. 7.9 Two dimensional implant profiles (a) portion of total dose as a function of lateral position for a mask (b) Equi-concentration contours for a 70 keV boron implant through a 1 μm slit.

7.5 THICKNESS OF MASKING

Now determine the thickness of a mask necessary to block the penetration of ion into the silicon, For example, photoresist is often used because impurity is performed at room temperature. Near the edge of the mask, the profile is dominated by the lateral straggle and given by a sum of point response function with the Gaussian equation (6.9). This shall mean the ion profile under the mask edge like the gate of MOSFET can be determined by the lateral straggle of the implant and by how far it moves during annealing. With decrease in device thickness, the straggle under the mask is the most interested region for study.

How thick should a mask be in order so that it can effectively block the transmission of ion through it? The thickness of the mask should be large enough such that the tail of the implant profile in the silicon is at some specified background concentration as shown in figure 7.10.

7.5 Thickness of Masking

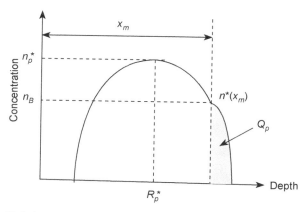

Fig. 7.10 Schematic of mask process showing a dose penetrating the mask of thickness x_m

The superscript * is use to identify the ranges and standard deviation in the masking material since they are in general different from those values in the silicon. The criterion for an efficient and effective masking has to follow equation (7.7).

$$n^*(x_m) = n_0^* \exp\left\{-\frac{(x_m - R_p^*)^2}{2\Delta R_p^{*2}}\right\} \qquad \ldots (7.12)$$

$$= \frac{Q_T}{\sqrt{2\pi}\Delta R_p^*} \exp\left\{-\frac{(x_m - R_p^*)^2}{2\Delta R_p^{*2}}\right\} \leq n_B \qquad \ldots (7.13)$$

Where n^* is the concentration at the far side of a mask of thickness x_m and C_B is the background concentration in the substrate. Setting $n^*(x_m) = n_B$ and solving for the mask thickness yields equation.

$$X_m = R_p^* + \Delta R_p^* \sqrt{2\ln\left(\frac{n_0^*}{n_B}\right)} = R_p^* + m\Delta R_p^* \qquad \ldots (7.14)$$

m is a parameter indicating that the thickness of the mask should be equal to the range plus multiple m times of the standard deviation in the masking material. Value m for different level of masking efficiency can be easily calculated from this equation.

If Q_P is the amount of dose that penetrates the mask then Q_P can be calculated using equation

$$Q_P = \frac{Q_T}{\sqrt{2\pi}\Delta R_p^*} \int_{x_m}^{\infty} \exp-\left(\frac{x - R_p^*}{\sqrt{2}\Delta R_p^*}\right)^2 dx \qquad \ldots (7.15)$$

This equation can be described as error function erf which is shown as

$$Q_p = \frac{Q_T}{2} erfc\left(\frac{X_m - R_p^*}{\sqrt{2}\Delta R_p^*}\right) \quad \ldots (7.16)$$

$$= \frac{Q_T}{2}\left[1 - erf\left(\frac{X_m - R_p^*}{\sqrt{2}\Delta R_p^*}\right)\right] \quad \ldots (7.17)$$

7.6 DOPING PROFILE OF ION IMPLANT

In general, the mask edge is not vertical or an angled implant is performed thus the numerical method must be used to calculate and show the 2D doping profile. There may be reason to use a high angle implant to introduce dopant underneath the MOS gate. In order to minimize short channel effect in small device, doping may be introduced below the tip extension region under the gate. This "halo" of doping is formed by high tilt implant under the edge of the gate. Tilted implant can cause shadowing effect due to the topography already on wafer. A TSUPREM V numerical simulation of the tilted implant near mask edge is shown in figure 7.11.

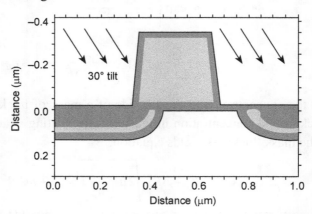

Fig. 7.11 Simulation result of 50KeV phosphorus implant at 30° tilt angle showing asymmetrical implant and shadow caused by gate Polysilicon Simulation done on TSUPREM V

To achieve symmetrical distribution for all devices, wafer often rotates during implant or is implanted in four separate rotations to completely avoid shadowing effect requires that the implant to be done at zero tilt angle.

To understand how the implant profile evolves in the time during subsequent annealing, one can compare the gaussian formulation from the implant distribution with that from a delta function distribution that has diffused.

7.7 Annealing

The solutions for both gaussian distribution in semi-medium are shown in equation below

$$n(x) = n_p \exp\left(-\frac{x - R_p^*}{\sqrt{2}\Delta R_p^*}\right)^2 \qquad \ldots (7.18)$$

$$= n(0)\exp\left(-\frac{x^2}{4Dt}\right) \qquad \ldots (7.19)$$

By comparing these solutions, one can see that the implanted gaussian profile with standard deviation R_p has the same form as initial delta function distribution that has diffused for an effective time temperature cycle of. Thus, the effect of additional time temperature cycle of annealing on the implanted gaussian distribution is expressed by equation.

$$n(x, t) = \frac{Q_T}{\sqrt{2\pi(\Delta R_p^2 + 2Dt)}} \exp\left(-\frac{(x - R_p)^2}{2(\Delta R_p^2 + 2Dt)}\right) \qquad \ldots (7.20)$$

From above equation it is clearly shown that a gaussian distribution remains as a gaussian distribution and it preserves its shape upon annealing in an infinite medium although its standard deviation or straggle about the peak concentration increases with the diffusion distance as shown in figure 7.12.

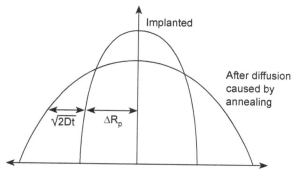

Fig. 7.12 Profile of ion implantation before and after annealing.

7.7 ANNEALING

After ion implantation, the wafer is usually so severely damaged that the electrical behavior is dominated by deep-level electron and hole traps that capture carriers and make the resistivity high. Annealing is required to repair lattice damage and put dopant atoms on substitutional sites. Figure 7.13 illustrates the effect of crystal structure before and after annealing.

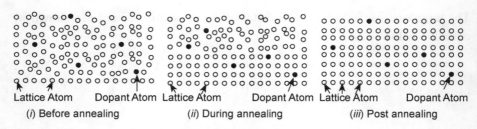

Fig. 7.13 Effect of RTA on crystal structure

The success of annealing is often measured by the fraction of dopant that is electrically active, as found experimentally using the Hall Effect technique. The Hall Effect measures an average effective doping level, which is integral over local doping densities and local mobilities evaluated per unit surface area

$$N_{\text{Hall}} = \frac{\left(\int_0^{x_j} \mu n\, dx\right)^2}{\int_0^{x_j} \mu^2 n\, dx} \qquad \ldots (7.21)$$

where μ denotes the mobility, n is the number of carriers, and x_j is the junction depth. If the mobility is not a strong function of depth, N_{Hall} measures the total number of electrically active dopant atoms. If annealing activates all of the implanted atoms, this value should be equivalent to the dose, Q_T.

7.7.1 Furnace Annealing

The annealing characteristics depend on the dopant type and the dose. There is a clear division between cases where the silicon has been amorphized and where it has been merely partially disordered. For amorphized silicon, re-growth proceeds via Solid Phase Epitaxy (SPE). The amorphous / crystalline interface migrates toward the surface at a fixed velocity that depends on temperature, doping, and crystal orientation. The SPE growth rate of amorphous silicon as a function of temperature for various crystal orientations is shown in figure 7.14. It should be noted that the activation energy for SPE is 2.3 eV, implying that the process involves bond breaking at the interface. The presence of impurities such as O, C, N and Ar impedes the regrowth process, as it is believed that these impurities bind to broken silicon bonds. Dopants such as B, P and As increase the regrowth rate (by a factor of 10 for concentrations in the regime of 10^{20} atoms/cm^3), presumably because substitutional impurities weaken bonds and increase the likelihood of broken bonds.

7.7 Annealing

The re-growth rate v is given by equation

$$v = A \exp\left(-\frac{2.3eV}{kT}\right) \qquad \ldots (7.22)$$

where A is an experimentally determined parameter. The re-growth rate can be enhanced by a factor of 10 for doping level characteristic of source and drain in MOS device. The re-growth at low temperature quickly eliminates all primary damage of the crystal in the amorphous region so that no anomalous dopant diffusion occurs. Most of the dopant atoms in the amorphous region are incorporated onto substitutional lattice sites during re-growth so that high levels of activation are possible even at low temperature.

If the implantation conditions are not sufficient to create an amorphous layer then the crystal lattice repair occurs by the generation and diffusion of point defects. The repair process has activation energy of about 5.0 eV and it requires temperature at least 900°C in order to remove all defects. Therefore, it is easier to repair fully amorphized layer than a partially damage layer.

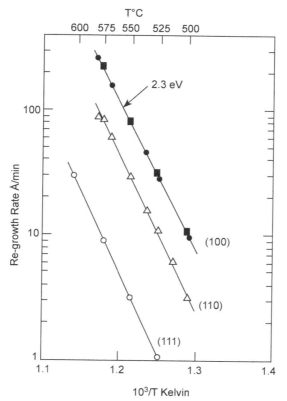

Fig. 7.14 Epitaxial growth rate of amorphous silicon as function of temperature for common crystal orientation

Annealing at high temperature causes competition between solid phase epitaxy SPE and local diffusive rearrangement that can lead to polysilicon formation. Thus, it is best to precede the high temperature step by low temperature regrowth. High temperature defect diffusion can then repair extended defects remaining after SPE. Annealing a partially damaged layer at low temperature can actually impair the process of lattice re-construction because the stable extended defects such as dislocation loop can be formed, which requires temperature of 1000°C to remove them.

7.7.2 Rapid Thermal Annealing (RTA)

Annealing has the objective to repair the damage with minimizing diffusion. Repairing crystal damage is a process with activation energy of 5.0 eV, while diffusion has activity energy ranges from 3.0 eV to 4.0 eV. Owing to different in activation energy, at sufficiently high temperature repair crystal lattice is faster than diffusion.

Furnace annealing is capable of supply high temperature but the practical steps require to load and to remove wafer without stressing them lead to minimum anneal time of 15 minutes. This time is much longer than the required time to repair crystal lattice at high temperature. Thus, diffusion is unavoided RTA is a method that covers various techniques of heating wafers for periods range from nanoseconds to 100s, allowing repair of crystal lattice with minimum diffusion.

Rapid thermal annealing can be divided into three classes, which are adiabatic, thermal flux, and isothermal annealing. In adiabatic annealing, the heating time is short, which is less than 1 μs. This type of annealing only affects a thin surface film. A high energy laser pulse can be used to melt the surface to a depth of less than 1.0 μm and surface crystallizes by liquid phase epitaxy with no defect left. Dopant diffusion in liquid state is very fast so that the final profile is roughly rectangular extending from the surface to the melt depth. By adjusting the pulse time and energy, a shallow junction can be obtained. But this method is not possible to preserve either doping profile or surface film. Thus, it is not generally used in VLSI circuit.

Thermal flux annealing method has annealing time ranges from 10^{-7}s to 1.0s. It is done by heating from one side of the wafer with a laser, electron beam, or flash lamp that provides a temperature gradient across the thickness of wafer. Generally, the surface is not melted but the surface damage can be repaired by SPE before any diffusion occurs. Figure 7.15 shows an arsenic profile annealed

7.7 Annealing

with scanned laser beam. Almost complete electrical activation is obtained without diffusion. Unfortunately, the rapid quenching from high temperature leaves many point defects which may condense to form dislocation, degrading the lifetime of minority carrier. In order to fabricate ultra-shallow junctions (discussed in more details in the next section), spike RTA technologies employing laser heating or other means expose the wafer to a high temperature for a fraction of a second.

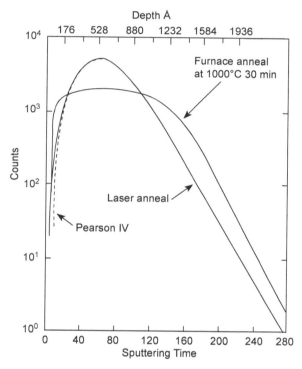

Fig. 7.15 Profile of As using conventional furnace and laser RTA method

The process time for isothermal annealing involved heating process, is longer than 1.0s. Isothermal annealing uses tungsten-halogen lamp or graphite resistive strip to heat the wafer from one or both sides of the wafer as shown in figure 7.16(a). Another method is furnace-based rapid thermal processing and the apparatus is shown schematically in figure 7.16(b). In this method, a thermal gradient is established by adjusting the power supplied to different zones of the bell jar, with the hottest zone on top. The sample is introduced rapidly into the zone to achieve RTA. A wafer temperature up to 1200°C and ramp rate of up to 150°C/sec can be achieved.

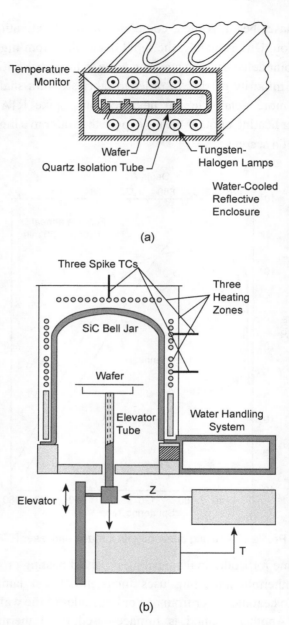

Fig. 7.16 (a) isothermal annealing system (b) Rapid thermal annealing system

7.8 SHALLOW JUNCTION FORMATION

In scaling horizontally to sub-micrometer and deep sub-micrometer dimensions, it is imperative to scale dopant profiles vertically. Junction depths on the order of tens of nanometers (nm) or less are required for deep sub-micrometer devices. These can be achieved by ion implantation using various methods.

7.8.1 Low Energy Implantation

For CMOS devices, shallow N^+ and P^+ layers are needed for the source and drain regions. As is heavy enough to form shallow N^+ layers with implantation energies permitted by commercial low energy ion implantation machines. However, in the case of B, the implanted atoms penetrate deeper, but the effective energy can be reduced by a factor of 11/49 by implanting the molecular ion BF_2, because upon impact BF_2^+ dissociates into atomic boron and fluorine. The extra fluorine atoms increase the lattice damage, thus minimizing channeling and facilitating annealing. New advances have enabled modern low energy ion implanters to deliver reasonably high ion current at energy below 1 keV. For ultra-shallow junction formation, laser doping and plasma doping are alternative techniques, but low energy beam-line ion implantation continues to be the main stream technology.

A profile can be moved closer to the surface by implanting through a surface film such as SiO_2. This shifts the profile by roughly the oxide thickness, but recoiled oxygen atoms can be problematic. This recoil effect can, however, be utilized if we dope by knocking dopant out of a deposited surface film using Si or inert gas implantation, a process termed ion mixing.

7.8.2 Tilted Ion Beam

If the wafer is tilted at a large angle relative to the ion beam, the vertical projected range can be decreased, as illustrated in figure 7.17. However, for large tilt angles, a significant fraction of the implanted ions is scattered out of

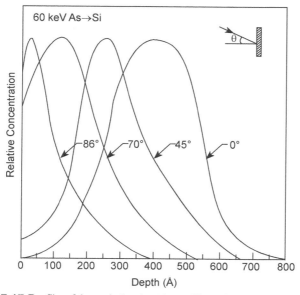

Fig. 7.17 Profile of Arsenic implant into silicon for various incident angles at 60 keV energy

the surface, so the effective dose is reduced. As a pragmatic technique, this is only useful when the wafer surface is not patterned because large tilt angles cause long shadows and asymmetries at mask edges.

7.8.3 Implanted Silicides and Polysilicon

The problem of forming a shallow layer can be circumvented if a surface layer is deposited and the dopant is subsequently diffused into the substrate from the surface layer. This is most often done when the surface film is to be used as an ohmic contact to the substrate. Carefully controlled diffusion can result in steep dopant profiles without damaging the silicon lattice. As shown in figure 7.18, dopant diffusion in polysilicon (or silicides) is generally much faster than insingle-crystal silicon, and so the implanted atoms soon become uniformly distributed in the surface thin film. Some of the dopant atoms diffuse into the substrate, consequently yielding a fairly abrupt profile like the one exhibited in figure 7.18. The small peak at the interface may be due to grain boundary

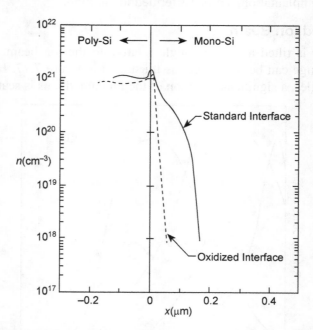

Fig. 7.18 Profile of Arsenic diffusion into silicon from a polysilicon source annealed at 950°C for 30 minute

segregation or impurities trapped at the interface. As shown by the dotted curve, the presence of a 25Å thick oxide layer between the polysilicon and the substrate is sufficient to block most of the diffusion. For silicides, there exists another option of implanting into the deposited metal film before the heat treatment that forms the silicide. If the implant is beneath the metal layer, the dopant atoms will be "snow ploughed" forward as the silicide forms, resulting in a steep dopant gradient near the interface. If the implant is inside the metal, it will segregate out at the moving silicide-silicon interface giving a very sharply peaked dopant distribution as well.

7.9 HIGH ENERGY IMPLANTATION

Implantation at MeV energy is frequently used in VLSI fabrication. The most common application is to form deep isolation regions among individual devices. It is also used to form tubs in CMOS structures (figure 7.19). High-energy ion implantation offers three ways in which the traditional epitaxial CMOS process can be improved. Firstly, the tubs can be implanted rather than diffused from the surface. In order to achieve a roughly uniform doped layer, a series of implants at different energies are required, accompanied by a short annealing step. Secondly, the structure can be improved further by retrograde doping of the tub, i.e. varying the implant doses such that the tub surface is less doped than the tub bottom. This method has the same advantages for the tub transistors as the use of epi-substrates has for the transistors outside the tub. Last but not least, the epi-substrate can be dispensed altogether by using a blanket high-energy implant in the first processing step. This will allow the formation of a buried, heavily-doped layer serving the same function as a heavily-doped substrate.

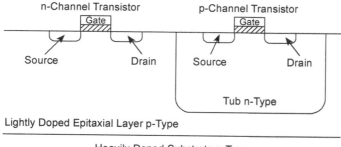

Fig 7.19 Cross section of an epi-substrate CMOS structure

7.10 BURIED INSULATOR

Devices can be fabricated in a thin silicon layer on an insulating substrate using two types of materials SOS (silicon-on-sapphire) and SOI (silicon-on-insulator). Both technologies have the advantage of increasing radiation hardness due to the reduced collection volume for charges generated by ionizing radiation. They also offer a compact way to isolate devices from each other to reduce parasitic capacitance and to eliminate latchup for CMOS circuits.

The SOS technology is more mature. High quality silicon epitaxial films have been successfully grown on sapphire wafers. However, sapphire substrates are very expensive, thereby limiting the use of SOS to demanding applications, such as military device.

One of the two common SOI techniques is to introduce a blanket buried oxide to isolate the device active region from the bulk wafer, a process called SIMOX (separation by implantation of oxygen). The principle behind the formation of a buried oxide layer is quite simple. If oxygen is implanted at a dose on the order of 10^{18} atoms/cm^2, there will be twice as many oxygen atoms as silicon atoms at the vicinity of the peak. Upon annealing, silicon dioxide will form.

The details of high dose oxygen implantation differ from traditional implantation steps in several ways. The goal is to maintain a surface layer of high-quality single-crystal silicon for device fabrication. Hence, the substrate is kept near 600°C during implantation so that self-annealing maintains the crystal integrity. Each incident oxygen ion sputters on the order of 0.1 silicon atom, and so the large number of implanted oxygen ion erodes many layers of silicon atoms. None the less, this is more than offset by the expansion in volume during the formation of oxide (44%). The net result is a slight swelling of the silicon surface. The implant profile also changes from a gaussian-like distribution to a flat-topped distribution, as depicted in figure 7.20. This is due to diffusion of oxygen to the silicon-oxide boundaries after oxygen saturates the substrate around the peak region. As implanted, the surface layer still contains a substantial amount of oxygen and much damage, albeit still single-crystal. Annealing is performed at a high temperature above 1300°C to cause a strong segregation of oxygen into the buried layer from both sides, consequently depleting the surface of almost all the implanted atoms, including impurities. This leaves a high-quality surface film containing very little oxygen and less than 10^9 dislocations/cm^2.

7.11 Summary

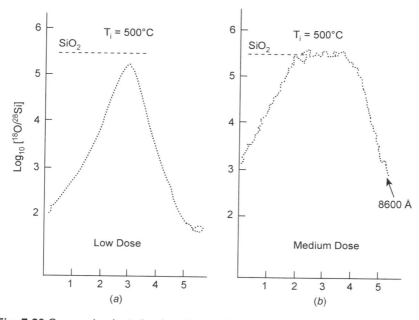

Fig. 7.20 Oxygen implantation into Si at 200 keV without annealing (a) Gaussian shape for low O_2 dope (b) become diffused toward the implant peak to form a stoichiometric buried SiO_2 layer at high annealing temperature and heavy oxygen dose

SOI substrates can be synthesized using wafer bonding and back etching. Using wafer bonding technology, two silicon wafers (one or both having a surface oxide layer) can be pasted together resulting in a $Si/SiO_2/Si$ structure. One side of the structure can be thinned by polishing to yield the required silicon layer thickness. The advantage of this technique is that the thickness of the oxide and silicon layers can be adjusted independently, but two wafers are required to make one SOI wafer thereby raising the cost.

7.11 SUMMARY

This chapter introduced the technology of ion implantation. The components of a modern ion implanter were described, and some limitations of ion implanting impurities were presented. Most implant profiles can be described by a gaussian distribution. Additional moments of the gaussian including skewness and kurtosis are sometimes used to better approximate experimentally observed profiles. After implantation the impurities must be annealed. The anneal step activates the impurities and repairs the implant damage. Different annealing recipes are called for depending on the amount of damage in the substrate. In silicon technologies, implantation can also be used to form buried insulators through high dose oxygen implantation. Finally the use of software to simulate implantation was covered.

PROBLEMS

1. A 40 keV implant of ^{11}B is done into bare silicon. The dose is 10^{12} cm^{-2}.
   (a) What is the depth of the peak of the implanted profile?
   (b) What is the concentration at this depth?
   (c) What is the concentration at a depth of 4000 Å (0.4 μm)?

2. Phosphorus is implanted into silicon. The implant parameters are a dose of 10^{16} cm^{-2} and an energy of 200 keV. Find the depth of the peak of the implant profile and its value at that depth.

3. Compare calculations of large profiles for implantation for 100 keV boron through 1400 Å of titanium silicide into silicon using the methods of R_p scaling and does matching. Neglect skewness and assume a dose of 10^{16} cm^{-2}.

4. A typical high-current implanter operates with an ion beam of 2 mA. How long would it take to implant a 140-mm-diameter wafer with O$^+$ to a dose of 1×10^{18} cm^{-2}?

5. The depth of the junction of the source/drain region in a MOSFET must be reduced as the gate length is scaled. It is highly desirable to produce low resistivity junctions thinner than 0.1 μm. Is this a significant problem for ion implantation? What are the major problems in forming these structures?

REFERENCES

1. J.W. Mayer, L. Erickson, and J.A. Davies, "Ion Implantation in Semiconductors, Silicon and Germanium," Academic Press, New York, (1970).

2. H. Ryssel and I. Ruge, "Handbook of Ion Implantation Technology," Wiley, New York, (1986).

3. G. Dearnaley, J.H. Freeman, R.S. Nelson, and J. Stephen, "Ion Implantation," New Holland Amsterdam, (1973).

References

4. T.C. Smith, "Wafer Cooling and Photoresist Masking Problems in Ion Implantation," Ion Implantation Equipment and Techniques, 11, Springer Series in Electrophysics, Springer-Verlag, New York, 196 (1983).

5. P. Burggraaf, "Resist Implant Problems: Some Solved, Others Not," *Semicond. Int.* 15:66 (1992).

6. J.F. Gibbons, "Ion Implantation in Semiconductors—Part I, Range Distribution Theory and Experiments," *Proc. IEEE* 56:295 (1968).

7. Y. Xia and C. Tan, "Four-parameter Formulae for the Electronic Stopping Cross-section of Low Energy Ions in Solids," *Nucl. Instrums, Methods* B, 13:100 (1986).

8. B. Smith, "Ion Implantation Range Data for Silicon and Germanium Device Technologies," Research Studies, Forest Grove (1977).

9. T.E. Seidel and S. M. Sze, "Ion Implantation," *VLSI Technology*, McGraw-Hill, New York, (1983).

10. J. Lindhard and M. Scharff, "Energy Dissipation by Ions in the kev Region," *Phys. Rev.* 124:128 (1961).

11. J. Lindhard, M. Scharff, and H. SchiOtt, "Range Concepts and Heavy Ion Ranges," *Mat. Fys. Med. Dan. Vidensk. Selsk* 33:14 (1963).

12. J.F. Gibbons, W.S. Johnson, and S.W. Mylroie, *"Projected Range Statistics,"* Dowden, Hutchinson, and Ross, Stroudsburg, PA, (1975).

13. W.P. Petersen, W. Fitchner, and E.H. Grosse, "Vectorized Monte Carlo Calculations for the Transport of Ions in Amorphous Targets," *IEEE Trans. Electron Dev.* 30:1011 (1983).

14. D.S. Gemmell, "Channeling and Related Effects in the Motion of Charged ParticlesThrough Crystals," *Rev. Mod. Phys.* 46:129 (1974).

15. N.L. Turner, "Effects of Planar Channeling Using Modern Ion Implant Equipment," *Solid State Technol.* 28:163 (1985).

16. G.H. Kinchi and R.S. Pease, "The Displacement of Atoms in Solids by Radiation," *Rep. Prog. Phys,* 18:1 (1955).

17. F.F. Morehead, B.L. Crowder, F. Eisen and L. Chadderton, "A Model for the Formation of Amorphous Silicon by Ion Implantation," *First International Conference on Ion Implantation*, Gordon and Breach, New York, (1971).

18. R.J. Schreutelkamp, J.S. Custer, J.R. Liefting, W.X. Lu, and F.W. Saris, "Preamorphization Damage in Ion-Implanted Silicon," *Mat. Sci. Rep.* 6:275 (1991).

19. T.Y. Tan, "Dislocation Nucleation Models from Point Defect Condensations in Silicon and Germanium," *Materials Res. Soc. Symp. Proc.* 2:163 (1981).

20. B.L. Crowder and R.S. Title, "The Distribution of Damage Produced by Ion Implantation of Silicon at Room Temperature," *Radiation Effects* 6:63 (1970).

21. T.E. Seidel, A.U. MacRae, F. Eisen and L. Chadderton, "The Isothermal Annealing of Boron Implanted Silicon," *First International Conference on Ion Implantation,* Gordon and Breach, New York, (1971).

22. S. Wolf and R.N. Tauber, "*Silicon Processing for the VLSI Era* vol.1," Lattice Press, Sunset Beach, CA, (1986).

23. M.I. Current and D.K. Sadana, "Materials Characterization for Ion Implantation," in *VLSI Electronics—Micro structure Science* 6, Academic Press, New York, (1983).

24. L. Csepergi, E.F. Kennedy, J.W. Mayer, and T.W. Sigmon, "Substrate Orientation Dependence of the Epitaxial Growth Rate for Si-implanted Amorphous Silicon," *J. Appl. Phys.* 49:3906 (1978).

25 J.A. Pals, S.D. Brotherton, A.H. van Ommen, and H.J. Ligthart, "Recent Developments in Ion Implantation in Silicon," *Mat. Sci. Eng.* B,4:87 (1989).

26. J. Gyulai, J. Ziegler, "Annealing and Activation", *Handbook of Ion Implantaion Technology,* Elsevier Science, Amsterdam (1992).

27. S. Radovanov, G. Angel, J. Cummings, and J. Buff , "Transport of Low Energy Ion Beam with Space Charge Compensation," 54th Annual Gaseous Electronics Conference, State College, PA, American Physical Society, (2001).

28. B. Thompson and M. Eacobacci, "Maximizing Hydrogen Pumping Speed in Cryopumps Without Compromising Safety," *Micro Mag.* May (2003).

29. K. Ohyu and T. Itoga, "Advantage of Fluorine Introduction in Boron Implanted Shallow P-n Junction Formation." *Jpn. J. Appl. Phys.* 29(3):457 (1989).

30 D. Lin and T. Rost, "The Impact of Fluorine on CMOS Channel Length and Shallow Junction Formation," *IEDM Technical Digest,* 843 (1993).

31. M.A. Foad, R. Webb, R. Smith, J. Matsuo, A. Al Bayati, T.S. Wang, and T. Cullis, "Shallow Junction Formation by Decaborane Molecular Ion Implantation," *J. Vacuum Sci. Technol. B: Microelectron Nanometer Struct.* 18(1):445-449 (2000).

32. A.S. Perel, W.K. Loizides, and W.E. Reynolds, "A Decaborane Ion Source for High Current Implantation," *Rev. Sci. Instrum.* 73(2 II):877 (2002).

33. M.A. Albano, V. Babaram, J.M. Poate, M. Sosnowski, and D.C. Jacobson, "Low Energy Implantation of Boron with Decaborane Ion," *Materials Research Society Symposium, Proceedings* 610:B3.6.1-B3.6.6 (2000).

34. C. Li, M.A. Albano, L. Gladczuk, and M. Sosnowski, "Characteristics of Ultra Shallow B Implantation with Decaborane," *Materials Research Society Symposium, Proceedings* 745: 235 (2002).

35. "Axcelis", "Imax High Dose, Low Energy Boron Cluster Implant Technology Added to Optima Platform," *Semicond. Fabtech* (2006).

36. D. Jacobson, T. Horsky, W. Krull, and B. Milgate, "Ultra-high Resolution Mass Spectroscopy of Boron Cluster Ions," *Nucl. Instrum. Methods, Phys. Res. B: Beam Interactions Mater. Atoms (1-2): August, 2005,* Ion Implantation Technology Proceedings of the 15th International Conference on Ion Implantation Technology, 406-410, (2005).

37. http://www.amat.com/products/Quantum.html?menuID=1_9_1.

38. K. Sridharan, S. Anders, M. Nastasi, K.C. Walter, A. Anders, O.R. Monteiro, and W. Ensinger, "Nonsemiconductor Applications," in *Handbook of Plasma Immersion Ion Implantation and Deposition*, Wiley, New York, (2000).

39. P.K. Chu, N.W. Cheung, C. Chan, B. Mizumo, and O.R. Monteiro, "Semiconductor Applications," *Handbook of Plasma Immersion Ion Implantation and Deposition,* Wiley, New York, (2000).

40. X.Y. Qian, N.W. Cheung, M.A. Lieberman, M.I. Current, P.K. Chu, W.L. Harrington, C.W. Magee, and E.M. Botnick, "Sub-100 nm p_/n Junction Formation Using Plasma Immersion Ion Implantation," *Nucl. Instrum. Methods Phys. Res. B: Beam Interactions Mater. Atoms 55(1-4):821 (1991).*

41. M. Takase and B. Mizuno, "New Doping Technology—Plasma Doping for Next Generation CMOS Process with Ultra Shallow Junction-LSI Yield and Surface Contamination Issues," *IEEE International Symposium on Semiconductor Manufacturing Conference, Proceedings,* B9-B11 (1997),

42. K.H. Weiner, P.G. Carey, A.M. McCarthy, and T.W. Sigmon, "Low-Temperature Fabrication of p-n Diodes with 300-angstrom Junction Depth," *IEEE Electron Dev. Lett.* 13(7):369-371 (1992).

43. K.H. Weiner, and A.M. McCarthy, "Fabrication of Sub-40-nm p-n Junctions for 0.18 μm MOS Device Applications Using a Cluster-Tool- Compatible, Nanosecond Thermal Doping Technique," *Proc. SPIE* 2091:63-70 (1994).

44. D.R. Myers and R.G. Wilson, "Alignment Effects on Implantation Profiles in Silicon," *Radiation Effects* 47:91 (1980).

45. M.C. Ozturk and J.J. Wortman, "Electrical Properties of Shallow P_n Junctions Formed by BF_2 Ion Implantation in Germanium Preamorphized Silicon," *Appl. Phys. Lett.* 52:281 (1988).

46. H. Ishiwara and S. Horita, "Formation of Shallow P_n Junctions by B-Implantation in Si Substrates with Amorphous Layers," *Jpn. J. Appl. Phys.* 24:568 (1985).

47. T.E. Seidel, D.J. Linscher, C.S. Pai, R.V. Knoell, D.M. Mather, and D.C. Johnson, "A Review of Rapid Thermal Annealing (RTA) of B, BF_2 and As Implanted into Silicon," *Nucl. Instrum. Methods B* 7/8:251 (1985).

48. T.O. Sedgwick, A.E. Michael, V.R. Deline, and S.A. Cohen, "Transient Boron Diffusion in Ion-implanted Crystalline and Amorphous Silicon," *J. Appl. Phys.* 63:1452 (1988).

49. G.S. Oehrlein, S.A. Cohen, and T.O. Sedgwick, "Diffusion of Phosphorus During Rapid Thermal Annealing of Ion Implanted Silicon," *Appl. Phys. Lett.* 45:417 (1984).

50. H. Metzner, G. Suzler, W. Seelinger, B. Ittermann, H.P. Frank, B. Fischer, K.H. Ergezinger, R. Dippel, E. Diehl, H.J. Stockmann, and H. Ackermann, "Bulk-Doping-Controlled Implant Site of Boron in Silicon," *Phys. Rev. B.* 42:11419 (1990).

51. L.C. Hopkins, T.E. Seidel, J.S. Williams, and J.C. Bean, "Enhanced Diffusion in Boron Implanted Silicon," *J. Electrochem. Soc.* 132:2035 (1985).

52. R.B. Fair, J.J. Wortman, and J. Liu, "Modeling Rapid Thermal Diffusion of Arsenic and Boron in Silicon," *J. Electrochem. Soc.* 131:2387 (1984).

53. A.E. Michael, W. Rausch, P.A. Ronsheim, and R.H. Kastl, "Rapid Annealing and the Anomalous Diffusion of Ion Implanted Boron," *Appl. Phys. Lett.* 50:416 (1987).

54. K.J. Reeson, "Fabrication of Buried Layers of SiO_2 and Si_3N_4 Using Ion Beam Synthesis," *Nucl. Instrum. Methods* B19-20:269 (1987).

55. K. Izumi, M. Doken, and H. Ariyoshi, "CMOS Devices Fabricated on Buried SiO_2 Layers Formed by Oxygen Implantation in Silicon," *Electron. Lett.* 14:593 (1978).

56. H.W. Lam, "SIMOX SOI for Integrated Circuit Fabrication," *IEEE Circuits Devices,* 3:6 (1987).

57. P.L.F. Hemment, E. Maydell-Ondrusz, K.G. Stevens, J.A. Kilner, and J. Butcher, "Oxygen Distributions in Synthesized SiO_2 Layers Formed by High Dose O⁻ Implantation into Silicon," *Vacuum,* 34:203 (1984).

References

58. G.F. Celler, P.L.F. Hemment, K.W. West, and J.M. Gibson, "Improved SOI Films by High Dose Oxygen Implantation and Lamp Annealing," in *Semiconductor-on-Insulator and Thin Film Transistor Technology, Mater. Res. Soc. Symp. Proc.* 53, Boston, (1986).

59. S. Cristoloveanu, S. Gardner, C. Jaussaud, J. Margail, A.J. Auberton Herve, and M. Bruel, "Silicon on Insulator Material Formed by Oxygen Ion Implantation and High Temperature Annealing: Carrier Transport, Oxygen Activation, and Interface Properties," *J. Appl. Phys.* 62: 2793 (1987).

60. M.I. Current and W.A. Keenan, "A Performance Survey of Production Ion Implanters," *Solid State Technol.* 28:139 (1985).

61. H. Glawischnig and K. Noack, "Ion Implantation System Concepts," *Ion Implantation Science and Technology*, Academic Press, Orlando, FL, (1984).

62. P. Burggraaf, "Equipment Generated Particles: Ion Implantation," *Semicond. Int.* 14(10):78 (1991).

8

Film Deposition: Dielectric, Polysilicon and Metallization

8.1 INTRODUCTION

Fabricating IC requires different kinds of thin films which can be classified into five groups: a) epitaxy layers, b) thermal oxides c) dielectric layers d) polycrystalline silicon e) metal films. The growth of epitaxial layer and thermal oxides were discussed in chapter 2 and chapter 3 respectively. Dielectric layers such as Silicon dioxide (SiO_2) and silicon nitride (Si_3N_4) are used for insulation between conducting layers as masks for diffusion and ion implantation, for covering doped films to prevent the loss of dopants as well as for passivation to protect devices from impurities, moisture, and scratches. Phosphorus-doped silicon dioxide, commonly referred to as P-glass or phosphosilicate glass (PSG), is especially useful as a passivation layer because it inhibits the diffusion of impurities (such as Na), and it softens and flows at 950°C to 1100°C to create a smooth topography that is beneficial for depositing metals. Borophosphosilicate glass (BPSG), formed by incorporating both boron and phosphorus into the glass, flows at even lower temperatures between 850°C and 950°C. The smaller phosphorus content in BPSG reduces the severity of aluminum corrosion in the presence of moisture. Si_3N_4 is a barrier to Na diffusion, is nearly impervious to moisture, and has a low oxidation rate. The local oxidation of silicon (LOCOS) process also uses Si_3N_4 as a mask. The patterned Si_3N_4 will prevent the underlying silicon from oxidation but leave the exposed silicon to be oxidized. Si_3N_4 is also used as the dielectric for DRAM MOS capacitors when it combines with SiO_2.

Polycrystalline silicon better known as poly-silicon is used as a gate electrode material in MOS devices, a conducting material for multi-level metallization and a contact material for devices having shallow junctions. Poly-silicon can be undoped or doped with elements such as As, P, or B to reduce the resistivity. The dopant can be incorporated in-situ during deposition, or later by diffusion or ion implantation. Polysilicon consisting of several percent oxygen is a semi-insulating material for circuit passivation.

Metallization is the final step in the wafer processing sequence. It is the process by which the components of IC's are interconnected by conductor material most common is aluminum. This process produces a thin-film metal layer that will serve as the required conductor pattern for the interconnection of the various components on the chip. Another use of metallization is to produce metalized areas called bonding pads around the periphery of the chip to produce metalized areas for the bonding of wire leads from the package to the chip. The bonding wires are typically 25 μm diameter gold wires, and the bonding pads are usually made to be around 100×100 μm^2 to accommodate fully the flattened ends of the bonding wires and to allow for some registration errors in the placement of the wires on the pads.

Here we will discuss processes in details some of them are already discussed in chapter 2 (CVD & MBE) that can be used for depositing thin films. These are very important set of process since all of the layers above the surface of the wafer must be deposited; processes used to deposit semiconducting and insulating layers often involve chemical reactions. This distinction, however, is changing now, generally the techniques used to deposit metals are physical i.e., they do not involve a chemical reaction. Section begins with the physical process of evaporation it also covers the second physical deposition process: sputtering. Sputtering has been used extensively in silicon technologies. It is able to deposit a wide range of alloys and compounds and has a good ability to cover surface topology. Then will cover chemical vapor deposition which has the best ability to cover surface topology of the three methods and produces the least substrate damage because the chemistry of each process is unique.

8.2 PHYSICAL VAPOR DEPOSITION (PVD)

Physical vapor deposition is a process in which the target material is first vaporized or sputtering then condensed on the substrate surface. There are no chemical reaction involve in the vicinity of substrate surface. In PVD the atoms or molecules in the vapor phase physically absorb on the surface of the substrate to form a solid film which is normally amorphous in nature and can be converted to crystalline form on proper annealing ambient. PVD occurred at a very low pressure so that very few gas collision occur with surface reaction occur very rapidly and very little rearrangement of atoms happens at the surface of wafer.

8.2 Physical Vapor Deposition (PVD)

Physical vapor deposition (PVD) technology fall into three typical classes.

(1) Evaporation

(2) Sputtering

(3) Molecular beam epitaxy (MBE).

We have already discussed MBE in detail in chapter 2 so here we will focus on Evaporation and sputtering methods.

8.2.1 Evaporation

Evaporation is one of the oldest techniques for depositing thin films. A vapor is first generated by evaporating a source material in a vacuum chamber and then transported from the source to the substrate and condensed to a solid film on the substrate surface as shown in figure 8.1. The wafers are loaded into a high vacuum chamber that is commonly pumped with either a diffusion pump or a cryo pump. Diffusion-pumped systems commonly have a cold trap to prevent the back streaming of pump oil vapors into the chamber. The charge, or material to be deposited, is loaded into a heated container called the crucible. It can be heated very simply by means of an embedded resistance heater through external power supply. As the material in the crucible becomes hot, the charge gives off a vapor. Since the pressure in the chamber is much less than 1 m.torr, the atoms of the vapor travel across the chamber in a straight line until they strike a surface, where they accumulate as a film. Evaporation systems may contain up to four crucibles to allow the deposition of multiple layers without

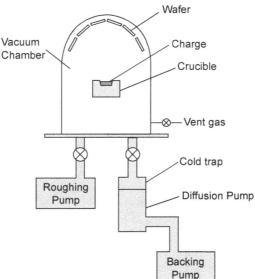

Fig. 8.1 Schematic of evaporation system containing diffusion pump and wafer containing chamber

breaking vacuum and may contain up to 24 wafers suspended in a frame above the crucibles. Furthermore, if an alloy is desired, multiple crucibles can be operated simultaneously. Mechanical shutters are used in front of the crucibles to provide start and stop the deposition abruptly.

8.2.2 Sputtering

Sputtering, unlike evaporation, is very well controlled and generally applicable to all materials such as metals, insulators, semiconductors, and alloys film deposition in microelectronic fabrication. It has a better step coverage than evaporation induces far less radiation damage than electron beam evaporation and is much better at producing layers of compound materials and alloys. These advantages made sputtering the metal deposition technique of choice for most silicon-based technologies until the advent of copper interconnect. Sputtering involves the ejection of surface atoms from an electrode surface by momentum transfer from the bombarding ions to the electrode surface atoms. The generated vapor of electrode material is then deposited on the substrate. A simple sputtering system, as shown in figure 8.2 is very similar to a simple reactive ion etch system a parallel-plate plasma reactor in a vacuum chamber. In a sputtering application, however, the plasma chamber must be arranged so that high energy ions strike a target containing the material to be deposited. The target material must be placed on the electrode with the maximum ion flux. To collect as many of these ejected atoms as possible, the cathode and anode in a simple sputtering system are closely spaced, often less than 15 cm. An inert gas is normally used to supply the chamber. The gas pressure in

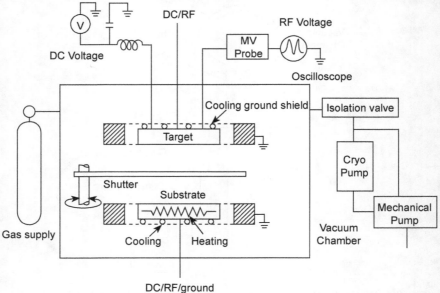

Fig. 8.2 Schematic diagram of typical sputtering system.

the chamber is held at about 0.1 torr. This results in a mean free path of order hundreds of μm. Due to the physical nature sputtering can be used for depositing a wide variety of materials. In the case of elemental metals, simple DC sputtering is usually favored due to its large sputter rates. When depositing insulating materials such as SiO_2, an RF plasma must be used.

8.3 CHEMICAL VAPOR DEPOSITION (CVD)

Chemical vapor deposition (CVD) has become extremely popular and is the preferred deposition method for a wide range of materials. Thermal CVD also forms the basis for most epitaxial growth in IC manufacturing. Modifications of simple thermal CVD processes provide alternate energy sources such as plasmas or optical excitation to drive the chemical reactions, allowing the deposition to occur at low temperature. The common CVD methods are

(1) Atmospheric-pressure Chemical Vapor Deposition (APCVD),
(2) Low-pressure Chemical Vapor Deposition (LPCVD), and
(3) Plasma-enhanced Chemical Vapor Deposition (PECVD).

A comparison between APCVD and LPCVD shows that the benefits of the low-pressure deposition processes are uniform step coverage, precise control of composition and structure, low-temperature processing, high enough deposition rates and throughput, and low processing costs. Furthermore, no carrier gases are required in LPCVD reducing particle contamination. The most serious disadvantage of LPCVD and APCVD is that their operating temperature is high, and PECVD is an appropriate method to solve this problem. Table 8.1 compares the characteristics and applications of the three CVD processes.

Table 8.1 Characteristics and applications of CVD processes

Process	Advantages	Disadvantages	Applications
APCVD (low T)	Simple reactor, fast deposition Low temperature	Poor step coverage, particle contamination Low throughput	Doped/undoped low T oxides
LPCVD	Excellent purity & uniformity, conformal step coverage, large wafer capacity, high throughput	High temperature, low deposition rate	Doped/ undoped high T oxides, silicon nitride, polysilicon, tungsten, WSi_2
PECVD	Low temperature, fast deposition, good step coverage	Chemical (eg. H_2) and particle contamination	Passivation (nitride), low T insulators over metals

Atmospheric-Pressure CVD (APCVD) reactors were the first to be used in the microelectronics industry. Operation at atmospheric pressure keeps reactor design simple and allows high deposition rates. However, the technique is susceptible to gas-phase reactions and the films typically exhibit poor step coverage. Since APCVD is generally conducted in the mass-transport-limited regime, the reactant flux to all parts of the every substrate in the reactor must be precisely controlled. Figure 8.3 shows the schematic of three typical APCVD reactors.

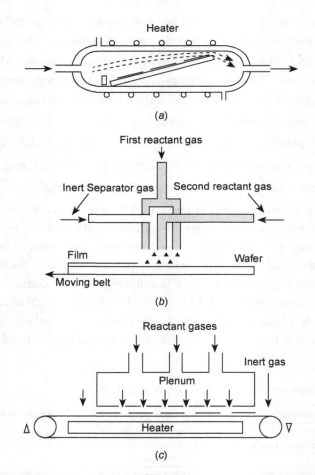

Fig. 8.3 Schematic of three typical APCVD reactors (a) horizontal tube (b) Gas injection type continuous process (c) Plenum type continuous processing

Figure 8.4 shows a typical commercial PECVD system. Rather than relying solely on thermal energy to sustain the chemical reactions, PECVD systems uses an RF-induced glow discharge to transfer energy into the reactant gases, allowing the substrate to remain at a lower temperature than that in APCVD and LPCVD. PECVD thus allows the deposition of films on substrates that do not have the

8.3 Chemical Vapor Deposition (CVD)

thermal stability. In addition, PECVD can enhance the deposition rate as compared with thermal reactions alone and can produce films of unique compositions and properties. However, the limited capacity, especially for large-diameter wafers, and possibility of particle contamination by loosely adhering deposits may be major concerns.

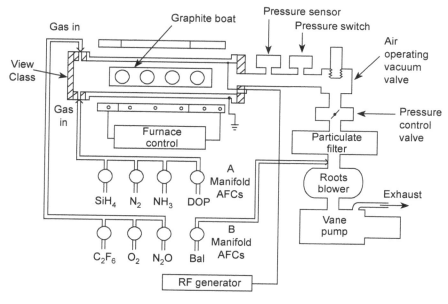

Fig. 8.4 Schematic diagram of a commercial PECVD system.

PECVD reactors are three general types: (1) parallel plate, (2) horizontal tube, and (3) single wafer. In the parallel plate reactor depicted in figure 8.5(a), the electrode spacing is typically 5 to 10 cm and the operating pressure is in the range of 0.1 to 5 torr. In spite of the simplicity, the parallel plate system suffers from low throughput for large-diameter wafers. Moreover, particulates flaking off from the walls or the upper electrode can fall on the horizontally positioned wafers.

A horizontal PECVD reactor resembles a hot-wall LPCVD system consisting of a long horizontal quartz tube that is radiantly heated. Special long rectangular graphite plates serve as both the electrodes to establish the plasma and holders of the wafers. The electrode configuration is designed to provide a uniform plasma environment for each wafer to ensure film uniformity. These vertically oriented graphite electrodes are stacked parallel to one another, side by side, with alternating plates serving as power and ground for the RF voltage. The plasma is formed in the space between each pair of plates.

A more recent PECVD reactor is the single-wafer design displayed in figure 8.5 (b). The reactor, which is load-locked, offers cassette-to-cassette operations and provides rapid radiant heating of each wafers as well as allowing in-situ monitoring of the film deposition. Wafers larger than 200 mm in diameter can be processed.

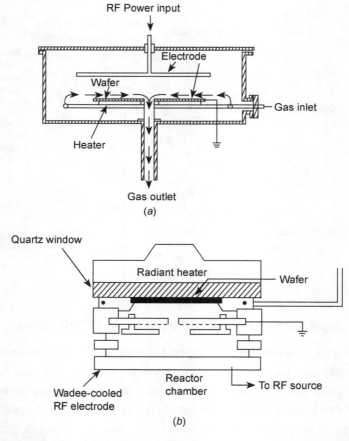

Fig. 8.5 Schematic diagram of PECVD reactors (a) Parallel plate type (b) single wafer type

Safety Issues

Most of the gases used for film deposition are toxic and these hazardous gases can also cause reactions with the vacuum pump oil. These hazardous gases can be divided into four general classes, pyrophoric (flammable or explosive), poisonous, corrosive, and dangerous combinations of gases. Gases commonly used in CVD are listed in Table 8.2. Gas combinations such as silane with halogens, silane with hydrogen, and oxygen with hydrogen will cause safety problems. In addition, silane reacts with air to form solid products causing

8.4 Silicon Dioxide

particle contamination in the gas lines. These particles can plug the pipes and perhaps create combustion.

Table 8.2 Gases commonly used in CVD

Gas	Formula	Hazards	Flammable Exposure limits in air (vol %)	Toxic limit (ppm)
Ammonia	NH_3	Toxic, corrosive	16-25	25
Argon	Ar	Inert	-	-
Arsine	AsH_3	Toxic	-	0.05
Diborane	B_2H_6	Toxic, flammable	1-98	0.1
Dichlorosilane	SiH_2Cl_2	Toxic, flammable	4-99	5
Hydrogen	H_2	Flammable	4-74	-
Hydrogen chloride	HCl	Toxic, corrosive	-	-
Nitrogen	N_2	Inert	-	-
Nitrogen oxide	N_2O	Oxidizer	-	-
Oxygen	O_2	Oxidizer	-	-
Phosphine	PH_3	Toxic, flammable	Pyrophoric	0.3
Silane	SiH_4	Toxic, flammable	Pyrophoric	0.5

8.4 SILICON DIOXIDE

Several deposition methods are used to produce silicon dioxide (SiO_2). Films can be deposited at lower than 500°C by reacting silane, dopant (for example phosphorus), and oxygen under reduced pressure or atmospheric pressure.

$$SiH_4 (g) + O_2 (g) \longrightarrow SiO_2 (s) + 2H_2 (g)$$
$$4PH_3 (g) + 5O_2 (g) \longrightarrow 2P_2O_5 (s) + 6H_2 (g)$$

The process can be conducted in an APCVD or LPCVD chamber. The main advantage of silane-oxygen reactions is the low deposition temperature allowing films to be deposited over aluminum metallization. The primary disadvantages are poor step coverage and high particle contamination caused by loosely adhering deposits on the reactor walls.

Silicon dioxide can be deposited at 650°C to 750°C in an LPCVD reactor by pyrolyzing tetra ethoxysilane, $Si(OC_2H_5)_4$. This compound, abbreviated TEOS, is vaporized from a liquid source.

The reaction is

$$Si(OC_2H_5)_4 (l) \longrightarrow SiO_2 (s) + \text{by-products} (g)$$

The by-products are organic and organosilicon compounds LPCVD TEOS it is often used to deposit the spacers beside the polysilicon gates. The process offers good uniformity and step coverage, but the high temperature limits its application on aluminum interconnects.

Silicon can also be deposited by LPCVD at about 900°C by reacting dichlorosilane with nitrous oxide:

$$SiCl_2H_2 \text{ (g)} + 2N_2O \text{ (g)} \longrightarrow SiO_2 \text{ (s)} + 2N_2 \text{ (g)} + 2HCl \text{ (g)}$$

This deposition technique provides excellent uniformity, and like LPCVD TEOS, it is employed to deposit insulating layers over polysilicon. However, this oxide is frequently contaminated with small amounts of chlorine that may react with polysilicon causing film cracking.

PECVD requires the control and optimization of the RF power density, frequency, and duty cycle in addition to the conditions similar to those of an LPCVD process such as gas composition, flow rate, deposition temperature, and pressure. Like the LPCVD process at low temperature, the PECVD process is surface-reaction-limited, and adequate substrate temperature control is thus necessary to ensure film thickness uniformity.

By reacting silane and oxygen or nitrous oxide in plasma, silicon dioxide films can be formed by the following reactions.

$$SiH_4 \text{ (g)} + O_2 \text{ (g)} \longrightarrow SiO_2 \text{ (s)} + 2H_2 \text{ (g)}$$
$$SiH_4 \text{ (g)} + 4N_2O \text{ (g)} \longrightarrow SiO_2 \text{ (s)} + 4N_2 \text{ (g)} + 2H_2O \text{ (g)}$$

Step Coverage and Reflow

The ability to cover surface topology is also called as step coverage. As the lateral dimensions of transistors decreases, the thickness of many layers remaines nearly constant. For the case, this step is the cross section of a contact etched through an insulating layer down to the substrate and over this scale of distance (<1 μm), the incoming material beam can be considered nondivergent. Three general types of step coverage are observed for deposited SiO_2, as schematically diagrammed in figure 8.6. A completely uniform or conformal step coverage depicted in Figure 8.6(a) results when reactants or reactive intermediates adsorb and then migrate promptly along the surface before reacting. When the reactants adsorb and react without significant surface migration, the deposition rate is proportional to the arrival angle of the gas molecules. Figure 8.6(b) illustrates an example in which the mean free path of the gas is much larger than the dimensions of the step.

The arrival angle in two dimensions at the top horizontal surface is 180°. At the top of the vertical step, the arrival angle is only 90° and so the film thickness is halved. Along the vertical walls, the arrival angle, φ, is determined by the width of the opening (w), and the distance from the top (z).

$$\varphi = \arctan(w/z) \qquad \ldots (8.1)$$

8.4 Silicon Dioxide

This type of step coverage is thin along the vertical walls and may have a crack at the bottom of the step caused by self-shadowing. Figure 8.6(c) depicts the situation where there is minimal surface mobility and the mean free path is short. Here the arrival angle at the top of the step is 270°, thus giving a thicker deposit. The arrival angle at the bottom of the step is only 90°, and so the film is thin. The thick cusp at the top of the step and the thin crevice at the bottom combine to give a re-entrant shape that is particularly difficult to cover with metal.

Doped oxides used as diffusion sources contain 5 to 15 % wt. of the dopant. Doped oxides for passivation or interlevel insulation contain 2 to 8% wt. phosphorus to prevent the diffusion of ionic impurities to the device. Phosphosilicate glass (PSG) used for the reflow process contains 6 to 8% wt. phosphorus. Oxides with lower phosphorus concentrations will not soften and flow, but higher phosphorus concentrations can give rise to deleterious effects for phosphorus can react with atmospheric moisture to form phosphoric acid which can consequently corrode the aluminum metallization. The addition of boron to PSG further reduces the reflow temperature without exacerbating

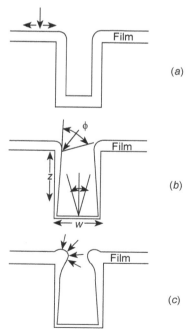

Fig. 8.6 Step coverage of deposited films. (a) Uniform coverage resulting from rapid surface migration (b) Nonconformal step coverage for long mean free path and no surface migration (c) Non conformal step coverage for short mean free path and no surface migration

this corrosion problem. Borophosphosilicate glass (BPSG) typically contains 4 to 6% wt. P and 1 to 4% wt. B. Poor step coverage of PSG or BPSG can be corrected by heating the samples until the glass softens and flow. PSG reflow is illustrated by the scanning electron micrographs shown in figure 8.7. Reflow is manifested by the progressive loss of detail.

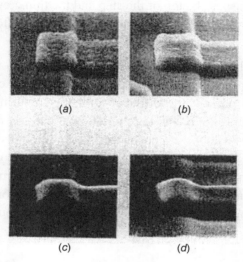

Fig. 8.7 SEM photographs (3200x) showing surfaces of 4.6 wt % P-glass annealed in steam at 1100°C for the following times (a) 0 min; (b) 20 min; (c) 40 min; (d) 60 min.

The step coverage of deposited oxides can be improved by planarization or etch-back techniques. Figure 8.8 illustrates the planarization process. Since

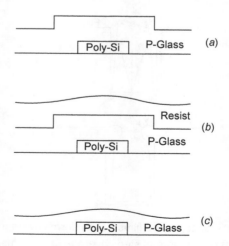

Fig 8.8 Schematic representation of the planarization process step covered with P-glass (b) Coated with resist (c) After leaving a smooth P-glass surface.

8.5 SILICON NITRIDE

the organic resist material has a low viscosity, reflow occurs during application or the subsequent bake. The sample is then plasma etched to remove all the organic coating and part of the PSG, as long as the etching conditions are selected to remove the organic material and PSG at equal rates.

8.5 SILICON NITRIDE

Stoichiometric silicon nitride (Si_3N_4) can be deposited at 700°C to 800°C at atmospheric pressure:

$$3SiH_4 (g) + 4NH_3 (g) \longrightarrow Si_3N_4 (s) + 12H_2 (g)$$

Using LPCVD, silicon nitride can be produced by reacting dichlorosilane and ammonia at temperature between 700°C and 800°C:

$$3SiCl_2H_2 (g) + 4NH_3 (g) \longrightarrow Si_3N_4 (s) + 6H_2 (g)$$

The reduced-pressure technique has the advantage of yielding good uniformity and higher wafer throughput.

Hydrogenated silicon nitride films can be deposited by reacting silane and ammonia or nitrogen in plasma at reduced temperature.

$$SiH_4 (g) + NH_3 (g) \longrightarrow SiN:H (s) + 3H_2 (g)$$
$$SiH_4 (g) + N_2 (g) \longrightarrow 2SiN:H (s) + 3H_2 (g)$$

Plasma-assisted deposition yields films at low temperature by reacting the gases in a glow discharge. Two plasma deposited materials, plasma deposited silicon nitride (SiNH) and plasma deposited silicon dioxide, are useful in VSLI. On account of the low deposition temperature 300°C to 350°C, plasma nitride can be deposited over the final device metallization. Plasma-deposited films contain large amounts of hydrogen (10 to 35 atomic). Hydrogen is bonded to silicon as Si-H, to nitrogen as N-H, and to oxygen as Si-OH and H_2O. Table 8.3 displays some of the properties of silicon nitride films fabricated using LPCVD and plasma-assisted deposition.

Table 8.3 Properties of silicon nitride films

Deposition	LPCVD	Plasma
Temperature(°C)	700-800	250-350
Composition	Si_3N_4(H)	SiN_xH_y
Si/N ratio	0.75	0.8-1.2
Atom % H	4-8	20-25
Refractive index	2.01	1.8-2.5
Density (g/cm³)	2.9-3.1	2.4-2.8
Dielectric constant	6-7	6-9

Deposition	LPCVD	Plasma
Resistivity (ohm-cm)	10^{16}	10^6-10^{15}
Dielectric strength (10^6 v/cm)	10	5
Energy gap (eV)	5	4-5
Stress (10^9 dyne/cm^2)	10 T	2C-5T

8.5.1 Locos Methods

The most straightforward way to produce a thick field oxide is by growing one before device fabrication, then etching holes in the oxide and fabricating the devices in these holes. This approach has two serious shortcomings. The first is the topology that is created. The step coverage for subsequent depositions will be poor and the photolithography will suffer. This is extremely serious if small features are to be printed. The second drawback is less obvious. On lightly doped substrates, a guard ring must be implanted to increase the parasitic threshold voltage. Unless very high energies are used, the implant must be done before the oxidation. Diffusion during oxidation may also be enhanced by point defects released during the oxidation process. Combined with alignment tolerance requirements, this will significantly reduce the density of the IC.

The isolation approach that has become the standard of silicon IC fabrication is local oxidation of silicon or LOCOS. Local oxidation is essentially an outgrowth of junction isolation and addresses both the isolation and the parasitic device formation concerns. A thin oxide is first grown and a layer of Si_3N_4 deposited on the wafer, usually by LPCVD. After the nitride is patterned, a field implant may be done to increase the threshold voltage of the parasitic MOSFET. Then the photoresist is stripped and the wafer is oxidized. The nitride acts as a barrier to the diffusion of the oxidant, preventing oxidation in selected regions of the silicon. A thin oxide will also be grown on top of the nitride. This is important because it limits the minimum nitride thickness to about 1000 Å and because the oxide must be removed before the nitride can be stripped after the field oxidation.

Since oxidation consumes 44% as much silicon as it grows, the resultant oxide is partially recessed and has a gradual step onto the field that is easy for photoresist and subsequent layers to cover. If the silicon is etched before the field implant the field oxide can be made fully recessed, resulting in a nearly planar surface. Figure 8.9 shows the growth of a local oxide along with a cross-sectional view of a completed LOCOS structure. The process leaves a characteristic bump on the surface, followed by a gradually narrowing oxide tail into the active area. The structure is called a bird's beak for obvious reasons. The bump, or bird's head, is particularly pronounced in recessed structures.

8.5 Silicon Nitride

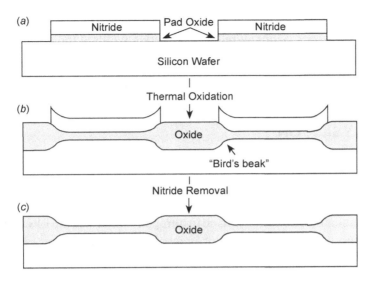

Fig. 8.9 Process sequence for local oxidation of silicon (LOCOS)

The purpose of the thin pad oxide layer under the nitride is to reduce the stress that occurs in the silicon substrate during oxidation. This stress is due to the mismatch of the thermal expansion coefficients of the substrate and the nitride and due to the volumetric increase of the growing oxide. At high temperature, viscous flow of the oxide greatly reduces the stress. A great deal of work has gone into optimizing the thicknesses of the oxide and nitride layers. If the stress exceeds the yield strength of silicon, it will generate dislocations in the substrate. A thicker pad oxide will lower the stress in the substrate. The minimum pad oxide thickness that can be tolerated without dislocation formation is about one-third the thickness of the nitride. This defect protection must be traded off against increased lateral encroachment of the oxide which occurs due to lateral diffusion of the oxidizing species through the pad oxide. A nitride to thermal pad oxide thickness ratio of 2.5:1 produces a lateral encroachment or bird's beak, approximately equal to the thickness of the field oxide.

One concern of the LOCOS process is the white ribbon or Kooi nitride effect. In this situation, a thermal oxynitride forms at the surface of the silicon under the edges of the nitride pad White ribbon is caused by the reaction of Si_3N_4 with the high temperature wet ambient to form NH_3, which diffuses to the Si/SiO_2 interface where it dissociates. When the effect is severe, the surface texture caused by these nitrides can be seen as a white ribbon around the edges of the active area. This defect leads to a reduced breakdown voltage in subsequent thermal oxides (such as gate oxides) in the active region. The existence of the bird's beak has two important consequences from a device standpoint often the active region defines the edge of the device in at least one direction. Then encroachment reduces the active width of the device, reducing the amount of current that a transistor will drive. A more subtle effect is due to the field doping. The field oxidation causes the field implant to diffuse into the edge of the active region.

8.6 POLYSILICON

Polysilicon is deposited by pyrolyzing silane between 575°C and 650°C in a low pressure reaction:

$$SiH_4 \text{ (g)} \longrightarrow Si \text{ (s)} + 2H_2 \text{ (g)}$$

Either pure silane or 20 to 30% silane in nitrogen is bled into the LPCVD system at a pressure of 0.2 to 1.0 torr. For practical use, a deposition rate of about 10 to 20 nm/min is required. The properties of the LPCVD polysilicon films are determined by the deposition pressure, silane concentration, deposition temperature, and dopant content.

Amorphous silicon can be prepared by the glow discharge decomposition of silane. Processing parameters such as deposition rate are affected by deposition variables such as the total pressure, reactant partial pressure, discharge frequency and power, electrode materials, gas species, reactor geometry, pumping speed, electrode spacing, and deposition temperature. The higher the deposition temperature and RF power, the higher is the deposition rate.

Polysilicon can be doped by adding phosphine, arsine, or diborane to the reactants (in-situ doping). Adding diborane causes a large increase in the deposition rate because diborane forms borane radicals, BH_3, that catalyze gas-phase reactions and increase the deposition rate. In contrast, adding phosphine or arsine causes a rapid reduction in the deposition rate, because phosphine or arsine is strongly adsorbed on the silicon substrate surface thereby inhibiting the dissociative chemisorption of SiH_4. Despite the poorer thickness uniformity across a wafer when dopants are incorporated, uniformity can be maintained by controlling precisely the flow of reactant gases around the samples.

Polysilicon can also be doped independently by other methods figure 8.10 shows the resistivity of polysilicon doped with phosphorus by diffusion, ion implantation, and in-situ doping. The dopant concentration in diffused polysilicon often exceeds the solid solubility limit, with the excess dopant atoms segregated at the grain boundaries. The high resistivity observed for lightly implanted polysilicon is caused by carrier traps at the grain boundaries. Once these traps are saturated with dopants, the resistivity decreases rapidly and approaches that for implanted single-crystal silicon.

8.7 Metallization

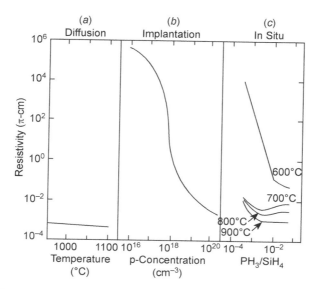

Fig. 8.10 Resistivity of phosphorous doped poly-silicon. (a) Diffusion: 1 hours at the indicated temperature (b) Implantation: 1 hour anneal at 1100°C. (c) In-situ: as deposited at 600°C and after a 30 minute anneal at the indicated temperature.

Poly-silicon can be oxidized in dry oxygen at temperatures between 900°C and 1000°C to form an insulator between the doped poly-silicon gate and other conducting layers. The resulting material, Semi-Insulating Poly-Silicon (SIPOS) is also employed as a passivating coating for high voltage devices.

8.7 METALLIZATION

A number of conductors such as Cu, Al and W etc, are used for fabrication of semiconductor devices. Metal with high conductivity is widely used for interconnection forming microelectronic circuit. Metallization is a process of adding a layer of metal on the surface of wafer. Metal such as Cu and Al are good conductors and they are widely used to make conducting lines to transport electrical power and signal. Miniature metal lines connect million of transistors made on the surface of semiconductor substrate. Metallization must have low resistivity for low power consumption and high integrated circuit speed, smooth surface for high resolution patterning process, high resistance to electro-migration to achieve high device reliability, and low film stress for good adhesion to underlying substrate. Other characteristics are stable mechanical and electrical properties during subsequent processing, good corrosion resistance, and relative receptivity to deposit and etch. It is important to reduce the resistance of the interconnection lines since integrated circuit device speed is closely related with RC constant time, which is proportional to the resistivity of the conductor used to form the metal line. Although Cu has lower resistivity than Al but technical difficulties such as adhesion, diffusion problem, and difficulties with dry etching etc have hampered Cu application in microelectronics for long time.

Aluminum has dominated metallization application since beginning of the semiconductor industry, In 1960s and 1970s, pure aluminum or aluminum siliconalloy were used as metal interconnection materials. By 1980s, when device dimension shrank, one layer of metal interconnect was no longer enough to route all the transistors and multi-layer interconnection became widely used. To increase the pack density, there must be near-vertical contact and via holes, which are too narrow for PVD of aluminum alloy to fill the via without voids. Thus, tungsten become a widely used material to fill contact and via holes and serves as the plug to connect different metal layers. Titanium and titanium nitride barrier/adhesion layers are deposited prior to tungsten deposition to prevent tungsten diffusion and peeling. Figure 8.11 illustrates a cross sectional view of a CMOS integrated circuit with aluminum interconnection and tungsten via plug. Borophosphosilicate glass (BPSG) is used as the insulating material separating the plugs.

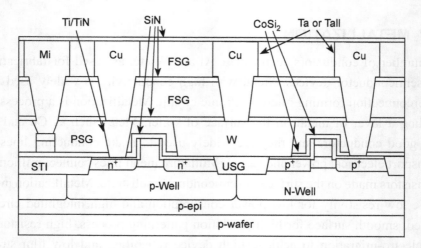

Fig. 8.11 Cross sectional view of a CMOS integrated circuit showing copper interconnection

8.8 METALLIZATION APPLICATION IN VLSI

For VLSI, metallization applications can be divided into three groups:
1. Gates for MOSFET
2. Ohmic Contacts
3. Interconnects.

Interconnection metallization interconnects thousands of MOSFETs or Bipolar devices using fine-line metal patterns. It is also same as gate metallization for MOSFET. All metallization directly in contact with semiconductor is called contact metallization. Polysilicon film is employed in the form of metallization used for gate and interconnection of MOS devices. Aluminium is used as the contact metal, on devices and as the second-level inter-connection to the outside world. Several new schemes for metallization have been suggested to produce ohmic contacts to a semiconductor. In several cases a multiple-layer structure involving a diffusion barrier has been recommended. Platinum silicide (PtSi) has been used as a Schottky barrier contact and also simply as an ohmic contact for deep junction. Titanium/platinum/gold or titanium/palladium/gold beam lead technology has been successful in providing high-reliability connection to the outside world. The applicability of any metallization scheme in VLSI depends on several requirements. However, the important requirements are the stability of the metallization throughout the IC fabrication process and its reliability during the actual use of the devices.

Ohmic Contacts

When a metal is deposited on the semiconductor a good ohmic contact should be formed. This is possible, if the deposition metal does not perturb device characteristics. Also die contact should be stable both electrically and mechanically.

Other important application of metallization is the top-level metal that provides a connection to the outside world. To reduce interconnection resistance and save area on a chip, multilevel metallization, as discussed in this section is also used. Metallization is also used to produce rectifying (Schottky barrier) contacts, guard rings, and diffusion barriers between reacting metallic films.

We have already stated the desired properties of metallization for ICs. None of the metals satisfies all the desired characteristics. Even Al, which has most of the desired properties suffers from a low melting point-limitation and electro migration as discussed above.

Poly-silicon has been used for gate metallization, for MOS devices. Recently, poly-silicon/refractory metal silicide bi-layers have replaced poly-silicon so that lower resistance an be achieved at the gale and interconnection level. By preserving the use of polysilicon as the "metal" in contact with the gate oxide, well known device characteristics and processes have been unaltered. The silicides of molybdenum ($MoSi_2$), tantalum ($TaSi_2$) and tungsten (WSi_2) have been used in the production of microprocessors and random-access memories. $TiSi_2$ and $CoSi_2$ have been suggested to replace $MoSi_2$, $TaSi_2$, and WSi_2 Aluminium and refractory metals tungsten and Mo are also being considered for the gate metal.

For contacts, Al has been the preferred metal for VLSI. However, for VLSI applications, several special factors such as shallower junctions, step coverage, electromigration (at higher current densities), and contact resistance can no longer be ignored. Therefore, several possible solutions to the contact problems in VLSI have been considered. These include use of

- Dilute Si-Ai alloy.
- Polysilicon layers between source, drain, or gate and top-level A1.
- Selectively deposited tungsten that is, deposited by CVD methods so that metal is deposited only on silicon and not on oxide.
- A diffusion barrier layer between silicon and Al, using a silicide, nitride, carbide, or their combination.

Use of self-aligned silicide, such as, PtSi, guarantees extremely good metallurgical contact between silicon and silicide. Silicides are also recommended in processes where shallow junctions and contacts are formed at the same time. The most important requirement of an effective metallization scheme in VLSI is that metal must adhere to the silicon in the windows and to the oxide that defines die window. In this respect, metals such as, Al, Ti, Ta, etc., that form oxides with a heat of formation higher than that of SiO_2 are the best. This is why titanium is the most commonly used adhesion promoter.

Although silicides are used for contact metallization, diffusion barrier is required to protect from interaction with Al which is used as the top metal. Aluminium interacts with most silicides in the temperature range of 200-500°C. Hence transition metal nitrides, carbides, and borides are used as a diffusion barrier between silicide (or Si) and Al due to their high chemical stability.

8.9 METTALIZATION CHOICES

In Metal Selection, the importance of interconnection metallization has been briefly discussed in the earlier introductory section, which is controlling the propagation delay by virtue of the resistance of interconnection line. The RC time constant of the line varies with silicon dioxide as the dielectric material follows equation (8.2).

$$RC = \frac{\rho_{Line}}{d_{Line}} \cdot \frac{L_{Line}\varepsilon_{or}}{d_{or}} \qquad \ldots (8.2)$$

Where ρ_{line} is the resistivity of the line material d_{Line} is the thickness of line L_{Line} is the length of the line d_{ox} is the thickness of oxide, and ε_{ox} permittivity of oxide.

The desired properties of the metallization for integrated circuit are as follows.
- Low resistivity
- Easy to form
- Easy to etch for pattern generation
- Should be stable in oxidizing ambient oxidizable
- Mechanical stability; good adhersion and low stress
- Surface smoothness
- Stability throughout processing, including high temperature sinter
- Dry and wet oxidation, gettering, phosphorous glass (or other material) passiviation, metallization, no reaction with final metal
- Should not contaminate device, wafer, or working equipment
- Good device characteristics and lifetime
- For window contact-low contact resistance, minimal junction penetration, low electromigration.

8.10 COPPER METALLIZATION

Instead of depending on chemical reaction to produce reacting species to form thin film like case of CVD, PVD can be used to deposit the film. PVD technique is generally more versatile than CVD method because it allows deposition of almost any material. Copper has lower resistivity (1.7 µΩ-cm) than aluminum copper alloy (2.9 to 3.3 µΩ-cm). It also has higher electron migration resistance because copper atoms are much heavier than aluminum and it has better reliability. Copper has always been an attractive and recommended choice for the metal interconnection in IC's industry because it can reduce power consumption and increase the speed of the IC's. Copper does have problem of adhesion with silicon dioxide and high diffusion rate in Si and SiO_2. Copper diffusion into Si can cause heavy metal contamination and lead to malfunction of the IC's. Thus, a barrier metal such as tantalum needs to be deposited before depositing copper. Copper is very hard to dry etch because copper-halogen compound has very low volatility. Copper also has issue of ansiotropicity due to lack of an effective dry etch method. It's an hindrance for the use of copper as common interconnects material for IC's fabrication.

8.11 ALUMINIUM METALLIZATION

Aluminum (Al) is the most commonly used material for the metallization of most IC's, discrete diodes, and transistors. The film thickness is as about 1 μm and conductor widths of about 2 to 25 μm are commonly used. The use of aluminum offers the following advantages:

- It has as relatively good conductivity.
- It is easy to deposit thin films of Al by vacuum evaporation.
- It has good adherence to the SiO_2 surface.
- Aluminum forms good mechanical bonds with Si by sintering at about 500°C or by alloying at the eutectic temperature of 577°C.
- Aluminum forms low-resistance, non-rectifying (that is, ohmic) contacts with p-type Si and with heavily doped n-type Si.
- It can be applied and patterned with a single deposition and etching process.

Aluminum has certain limitations.

1. During packaging operation if temperature goes too high, says 600°C, or if there is over heating due to current surge, Al can fuse and can penetrate through the oxide to the Si and may cause short circuit in the connection. By providing, adequate process control and testing, such failures can be minimized.
2. The silicon chip is usually mounted in the package by a gold perform or die backing that alloys with the silicon. Gold lead wires have been bonded to the aluminium film bonding pads on the chip, since package lead are usually gold plated. At elevated temperatures, a reaction between the metal of such systems causes formation of intermetallic compounds, known as the purple plague. Purple plague is one of six phases that can occur when gold and aluminium inter-diffuse. Because of dissimilar rate of diffusion of gold and aluminium, voids normally occur in the form of the purple plague. These voids may result in weakened bonds,

8.11 Aluminium Metallization

resistive bonds or catastrophic failure. The problem is generally solved by using aluminium lead wires, or another metal system, in circuits that will be subjected so elevated temperatures. One method is to deposit gold over an under layer of chromium. The chromium acts as a diffusion barrier to the gold and also adheres well to both oxide and gold. Gold has poor adhesion to oxide because it does not oxide itself. However, the chromium-gold process is comparatively expensive, and it has an uncontrollable reaction with silicon during alloying.

3. Aluminium suffers from electromigration which can cause considerable material transport in metals. It occurs because of the enhanced and directional mobility of atoms caused by the direct influence of the electric field and the collision of electrons with atoms, which leads to momentum transfer. In thin-film conductors that carry sufficient current density during device operations, the mode of material transport can occur at much lower temperature (compared to bulk metals) because of the presence of grain boundaries, dislocations and point defects that aid the material transport. Electromigration-induced failure is the most important mode of failure in Al lines.

Aluminum is the forth most conductive metal, after silver of resistivity 1.6 $\mu\Omega$cm element with resistivity of 2.65 $\mu\Omega$cm, and gold of resistivity of 2.2 $\mu\Omega$cm copper of resistivity of 1.7 $\mu\Omega$cm. Aluminum can be easily dry etched than the other three elements to form tiny metal interconnection lines. Both CVD and PVD processes can be used to deposit aluminum. PVD aluminum has higher quality and lower resistivity. PVD is a more popular method in microelectronic industry. Thermal evaporation, electron beam evaporation, and plasma sputtering can be used for aluminum PVD. Magnetron sputtering deposition is the most commonly used PVD process for aluminum alloy deposition in advanced fabrication. Aluminum CVD normally is a thermal CVD process with an aluminum organic compound such as dimethyl laluminium hydride (DMAH) $Al[(CH_3)_2H]$ with aluminum as the precursor. Aluminum interdiffuses into Si to form aluminum spikes. Aluminum spikes can punctual through the doped drain/source junction causing shorting to substrate silicon. The effect is called junction spiking as illustrated in figure 8.12. This problem can be solved by adding 1% of silicon to aluminum to form alloy instead of pure aluminum. Thermal annealing at 400°C forms Si-Al alloy at the silicon-aluminum interface that helps to prevent aluminum silicon inter diffusion causing junction spike.

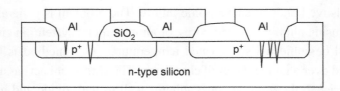

Fig. 8.12 Illustration of junction spiking caused by aluminum diffusion

Metallic aluminum is a polycrystalline material, which contains many small monocrystalline grains. When electric current flows through an aluminum line, a stream of electrons constantly bombards the grains some smaller grains start to move down just like the rock at the bottom of stream moving down during flood season. This effect is called electromigration. The illustration is shown in figure 8.13.

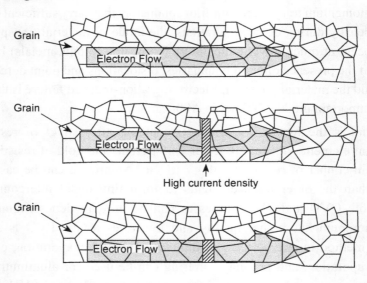

Fig. 8.13 Illustration of electromigration

Electromigration can cause serious problem for aluminum lines. When some grains begin to move due to electron bombardment, they damage the metal line. At some points, they cause higher current density at these points. This aggravates the electron bombardment and causes more aluminum grain.

High current and high resistance would generate heat and eventually cause breaking of aluminum line. Thus, electromigration can affect the reliability of microelectronic devices. Adding a small percentage of copper 0.5% wt. to aluminum can significantly improve the resistance of aluminum migration. This is because the copper atom is large and it can hold aluminum grains preventing migration due to electron bombardment.

8.12 METALLIZATION PROCESSES

Metallization process can be classified info two types.
1. CVD and
2. Physical Vapor Deposition

CVD offers three important advantages:
- Excellent step coverage
- Large throughput
- Low-temperature processing

The basic physical vapor deposition methods are :
- Evaporation
- Sputtering

Both these methods have three identical steps.
1. Converting the condensed phase (generally a solid) into a gaseous or vapor phase.
2. Transporting the gaseous phase from the source to the substrate.
3. Condensing the gaseous source on the substrate.

In both methods the substrate is away from the source.

In cases where a compound, such as silicide, nitride, or carbide, is deposited one of the components is as gas and the deposition process is termed reactive evaporation or sputtering.

8.13 DEPOSITION METHODS

In the evaporation method, which is the simplest a film is deposited by the condensation of the vapor on the substrate. The substrate is maintained at a lower temperature than that of the vapor. All metals vaporize when heated to sufficiently high temperatures. Several methods of heating are employed to attain these temperatures. For Al deposition, resistive, inductive (RF), electron bombardment, electron-gun or laser heating can be employed. For refractory metals, electron-gun is very common. Resistive heating provides low throughput. Electron-gun cause radiation damage, but by heat treatment it can be annealed out. This method is advantageous because the evaporations take place at pressure considerably lower than sputtering pressure. This makes the gas entrapment in the negligible. RF heating of the evaporating source could prove to be the best compromise in providing large throughput, clean environment, and minimal levels of radiation damage.

In sputtering deposition method the target material is bombarded by energetic ions to release some atoms. These atoms are then condensed on the substrate to form a film. Sputtering, unlike evaporation is very well controlled and is generally applicable to all materials metals, alloys, semiconductors and insulators. RF-DC and DC-magnetron sputtering can be used for metal deposition. Alloy-film deposition by sputtering from an alloy target is possible because the composition of the film is locked to the composition of the target. This is true even when there is considerable difference between the sputtering rates of the alloy components. Alloys can also be deposited with excellent control of composition by use of individual component targets. In certain cases, the compounds can be deposited by sputtering the metal in a reactive environment. Thus gases such as methane, ammonia, or nitrogen, and diborane can be used in the sputtering chamber to deposit carbide, nitride, and boride, respectively. This technique is called reactive sputtering. Sputtering is carried out at relatively high pressures (0.1 to 1 pascal or Pa). Because gas ions are the bombarding species, the films usually end up including small amount of gas. The trapped gases cause stress changes. Sputtering is a physical process in which the deposited film is also exposed to ion bombardment. Such ion bombardment causes sputtering damage, which leads to unwanted charges and internal electric fields that affect device proxide. However such damages can be annealed out at relatively low temperatures. (<500°C), unless the damage is so severe as to cause an irreversible breakdown of the gate dielectric.

8.14 DEPOSITION APPARATUS

The metallization is usually done in vacuum chambers. A mechanical pump can reduce the pressure to about 10 to 0.1 Pa. Such pressure may be sufficient for LPCVD. An oil-diffusion pump can bring the pressure down to 10^{-5} Pa and with the help of a liquid nitrogen trap as low as 10^{-7} Pa. A turbo molecular pump can bring the pressure down to 10^{-8}-10^{-9} Pa. Such pumps are oil-free and are useful in molecular-beam epitaxy where oil contamination must be avoided. Besides the pumping system, pressure gauges and controls, residual gas analyzers, temperature sensors, ability to clean the surface of the wafers by back sputtering, contamination control, and gas manifolds, and the use of automation should be evaluated.

As typical high-vacuum evaporation apparatus is shown in the figure 8.14 below.

8.14 Deposition Apparatus

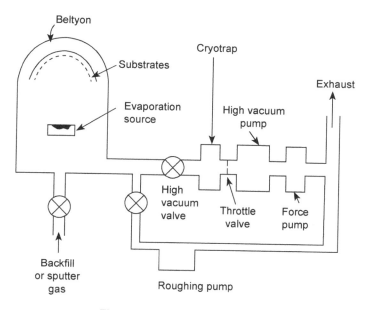

Fig. 8.14 Metallization Process

The apparatus consists of a hell jar, which is a stainless-steel cylindrical vessel closed at the top and sealed at the base by a gasket. Beginning at atmospheric pressure the jar is evacuated by a roughing pump, such as a mechanical rotary-van pump reducing pressure to about 20 Pa or a combination mechanical pump and liquid-nitrogen-cooled molecular pump (reducing pressure low about 0.5 Pa). At the appropriate pressure, the jar is opened to a high-vacuum pumping system that continues to reduce the pressure. The high-vacuum, pumping system may consist of a liquid nitrogen-cooled trap and an oil-diffusion pump, a trap and a turbomolecular pump, or a trap and a closed-cycle helium refrigerator cryopump. The cryopump acts as a trap and must be regenerated periodically, the turbomolecular and diffusion pumps act as transfer pumps, expelling their gas to a fore pump. The high vacuum pumping system brings the jar to a low pressure that is tolerable for the deposition process.

All components in the chamber are chemically cleaned and dried. Freedom from sodium contamination is vital when coating MOS devices.

The sputtering system operates with about 1 Pa of argon pressure during film deposition. For sputtering, a throttle valve should be placed between the trap and the high-vacuum pumping system. The argon gas pressure can to be maintained by reducing the effective pumping speed of the high-vacuum pump, while the full pumping speed of the trap for water vapor is utilized. Water vapor and oxygen are detrimental to film quality at background pressures of about 10^{-2} Pa.

The use of thickness monitors is common in evaporation and sputtering deposition. This is necessary for controlling the thickness of the film, because thinner film can cause excess current density and excessive thickness can lead to difficulties in etching.

Metallization Patterning

Once the thin-film metallization has been done the film must be patterned to produce the required interconnection and bonding pad configuration. This is done by a photolithographic process of the same type that is used for producing patterns in SiO_2 layers. Aluminium can be etched by a number of acid and base solutions including HCl, H_3PO_4, KOH, and NaOH. The most commonly used aluminium etchant is phosphoric acid with the addition of small amounts of HNO_3 and CH_3COOH, to result a moderate etch rate of about 1 μm/m at 50°C. Plasma etching can also be used with aluminium.

8.15 LIFTOFF PROCESS

The liftoff process is an alternative metallization patterning technique. In this process a positive photoresist is spun on the wafer and patterned using the standard photolithographic process. Then the metallization thin film is deposited on top of the remaining photoresist. The wafers are then immersed in suitable solvent such as acetone and at the same time subjected to ultrasonic agitation. This causes swelling and dissolution of the photoresist. As the photoresist comes off it liftsoff the metallization on top of it, for the liftoff process to work, the metallization film thickness must generally be somewhat less than the photoresist thickness. This process can, however produce a very fine line-width metallization pattern, even with metallization thickness that are greater than the line width.

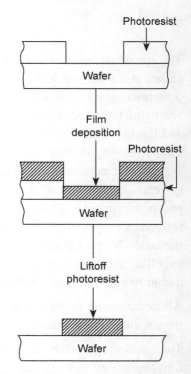

Fig. 8.15 Interconnection formation by additive metal liftoff

The additive or liftoff process is shown in figure 8.15 can also be used, in which the substrate is first covered with a photoresist patterned with openings where the final material is to appear. The thin film layer is deposited over the surface of the wafer. Any material deposited on top of the photoresist layer will be removed with the resist, leaving the patterned material on the substrate. For liftoff to work properly, there must be a very thin region or a gap between the upper and lower films, otherwise tearing and incomplete liftoff will occur.

8.16 MULTILEVEL METALLIZATION

A single layer of metal supply does not provide sufficient capability to fully interconnect complex VLSI chips. Many processes now use two or three levels of polysilicon, as well as several levels of metallization, in order to ensure the ability of making wires and provide adequate power distribution.

Silicon and GaAs technologies have taken very different approaches in making connection between the contact and the bonding pad. In large part, this is because the tasks for which these technologies have been designed are so different. Most silicon technologies have been designed to achieve high levels of integration. Many GaAs technologies have been optimized for high speed analog operation, with only a secondary emphasis on density. These technologies often simply place a single layer of gold on top of the Ni to reduce the interconnect resistance. This layer is deposited directly on top of the wafer after the transistors have been fabricated. The most important criterion is that the interconnect have a controlled characteristic impedance that is matched to the device input and output impedance. As the number of transistors on digital GaAs circuits has increased however, digital GaAs-based technologies have had to use multiple layers of interconnect. In doing so, they have sometimes grafted the same metallization approaches that have been used in silicon technologies for many years onto the basic MESFET technology. For that reason, we will approach the metallization process primarily as technology independent.

Multilevel metallizationcan be divided into three categories.
1. Basic multilevel Metallization
2. Planarized Metallization
3. Low dielectric constant interlevel Dielectrics

Basic Multilevel Metallization

A multilevel metal system is shown in figure 8.16. standard processing is used through the deposition and patterning of the first level of metal. An interlevel dielectric, consisting of CVD.

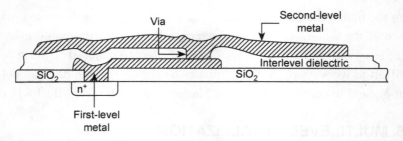

Fig. 8.16 Basic two level metallization

or sputtered SiO_2, or a plastic like material called polymide, is then deposited over the first metal layer. The dielectric layer must provide good step coverage and should help smooth the topology. In addition, the layer must be free of pinholes and be a good insulator. Next, vias are opened the dielectric layer, and the second layer of metallization is deposited and are patterned.

Most millimeter microwave ICs (MMICs) often have only a small number of transistors. Many discrete microwave components are also manufactured. In these technologies, wire density is not typically an issue. One layer of metallization is used to interconnect the devices. The Schottky gate metal is also often used as a first layer for parallel-plate capacitors and, where necessary, the Ni/AuGe serves as the second layer of metallization.

In modern silicon technologies a distinction is drawn between global or "true" interconnect and local interconnect. Although the resistivity of aluminum is low enough that long runs do not degrade performance in most circumstances, the same cannot be said of polysilicon. Its resistivity is typically 10^{-4} Ω.cm. Silicides, which can be run directly on top of the polysilicon to shunt the poly resistance, still have resistivities that are much larger than aluminum (Table 8.4).

Table 8.4 Interconnection materials property

Material	Melting Point (°C)	bulk resistivity (µΩ.cm)
Au	1064	2.2
Al	660	2.7
Cu	1083	1.7
W	3410	5.7
$TiSi_2$	1540	15
WSi_2	2165	40

Planarized Metallization

The topology that results from the simple multilayer interconnect process of figure 8.16 simply cannot be utilized in submicron processes because of the depth of field limitations in the lithographic processes. The Chemical Mechanical Planarization (CMP) process is used to achieve highly planar layers. In the process flow shown in figure 8.17 via filling technique is used to form the via between metal layers.

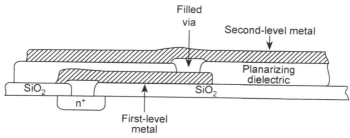

Fig. 8.17 Additional process step to achieve more planar structure

Tungsten is commonly used as the via metallization. The dielectric deposition, metallization, and CMP processes are repeated until the desired number of levels of interconnections is achieved. Integrated circuits with six levels of metal have been successfully fabricated using similar processes.

Low Dielectric Constant Interval Dielectrics

Propagation delay associated with interconnections is a critical issue in high performance microprocessors, as well as other integrated circuits. The RC product associated with these interconnections can be decreased by reducing either the resistance or capacitance or both.

8.17 CHARACTERISTICS OF METAL THIN FILM

Conducting film usually has polycrystalline structure. The conductivity and reflectivity of a metal are related to the grain size. Normally larger grain size has greater conductivity and inferior reflectivity. For higher temperature deposition, normally it has higher mobility to deposit atoms on the substrate surface and formed large grain in the deposited film. In the sub-section, we shall discuss some characteristics of metal film that cover the thickness of the film, the uniformity, the stress, the reflectivity, and the RC constant of the film.

Thickness of Metal Film

The thickness measurement for metal thin film is quite different from that of dielectric thin film. It is difficult to directly and precisely measure the thickness of an opaque thin film such as aluminum, titanium, titanium nitride, and copper, which usually needs to be performed on test wafer in a destructive way until the introduction of the acoustic measurement method. The metal film needs to be removed and its thickness either measured by Scanning Electron Microscope (SEM) or by measuring the step height with profilometer. Energetic electron beam scans across the metal film creates secondary electron emission from the metal sample. By measuring the intensity of the secondary electron emission, the thickness can be known from the image of secondary electron emission. It is also known that the different metal will have different rate of emission of secondary electron. SEM method can also detect void in the metal film.

Profilometer measurement can provide information pertaining the thickness and uniformity for film thicker than 1,000 Å. The pattern of metal is required to be deposited before it is being measured by stylus probe of profilometer as shown in figure 8.18.

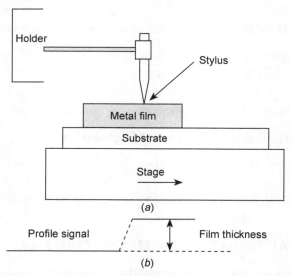

Fig. 8.18 (a) Schematic of stylus profilometer and (b) the profile of thickness

The metal pattern is done by the depositing a layer of metal on silicon substrate. It is then followed lithography process with a metal pattern mask to form a metal pattern on the photoresist. After development and etch processes, the metallic pattern is remained on Si-substrate. This metallic pattern is then measured with profilometer to determine its thickness. Ultrathin titanium nitride from 50 to 100Å is almost transparent. Its thickness can be measured with reflecto-spectrometer. The four-point probe is also commonly used to indirectly monitor the metal film thickness by assuming the resistivity of the metal film is constant throughout the wafer.

Uniformity of Metal Film

The non-uniformity of the metal film can be measured by measuring the sheet resistance and reflectivity at multiply locations on the wafer in the pattern illustrated in figure 8.19.

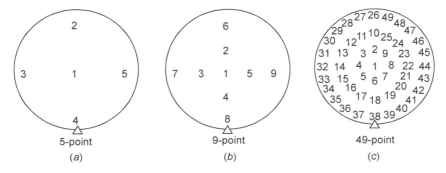

Fig. 8.19 The mapping pattern of wafer with (a) point measurement, (b) 9-point measurement, and (c) 49-point measurement

The more measurement points taken the more precision can be achieved. In the industry, 5-point and 9-point measurements are commonly used to save cost and time. The 49-point and three sigma (3σ) standard deviations is the most common defined criteria for qualification process of semiconductor industry.

Stress of Metal Film

Stress is caused by material mismatch between the film and the substrate. There are two type stresses namely compressive and tensile types. If stress is too high irrespective of the types, it can cause serious problem such as hillock and crack as shown in figure 8.20.

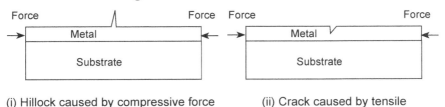

Fig. 8.20 Hillock and crack caused by high stress

There are two type of stresses

Intrinsic stress and thermal stress. Intrinsic stress is caused by film density. It is determined by ion bombardment in the plasma sputtering deposition process. When the atoms on the wafer surface are bombarded by the energetic ions from the plasma with much force, they are squeezed densely together while forming the film. This type of film would expand but it is being compressed by the substrate and this type of stress is the compressive type. Higher deposition temperature increases the mobility of the atoms, which in turn increases the film density and causes less tensile stress. Intrinsic stress is also related with wafer temperature change and different thermal expansion coefficient of the thin film and substrate. The thermal expansion coefficient of aluminum is 23.6 × $10^{-6} k^{-1}$ and thermal expansion coefficient of silicon is $2.6 \times 10^{-6} k^{-1}$. When an aluminum film is deposited at high temperature of 250 °C on Si-substrate at high temperature, upon cooling down aluminum shrinks more than silicon. This would result tensile stress on aluminum thin film by the silicon substrate.

Reflectivity of Metal Film

Reflectivity is a significant property of metal thin film. For stable metallization process, the reflectivity of the deposited film should be constant. The change of reflectivity is a sign of process drift. Reflectivity is a function of the film grain size and surface smoothness. Normally the larger grain size means lower reflectivity. The smoother the metal surface, the higher will be the reflectivity.

Reflectivity is important to photolithography process because it can cause a standing wave effect due to interference between incoming light and reflected light causing wavy grooved on the sidewall of the photoresist stack from periodic over exposure and underexposure. Anti-reflectant is normally coated to prevent this effect during pattern process especially the aluminum patterning that has very high reflectivity (180 to 200% relative to silicon). You may recall what has been discussed pertaining to this issue in lithography class.

Example:

A 0.25 µm metal line is 500 µm long. It is on top of 0.5 µm of SiO_2, and there are two more identical lines, one on each side. The line-to-line spacing is 0.25 µm. This space is also filled with SiO_2. Neglecting fringing effects, calculate wire-to-wire and wire-to-substrate capacitances for 0.40 µm thick Cu and 0.65 µm thick Al/Cu.

Solution

In either case, the wire-to-substrate capacitance is

$$C_{w-s} = \frac{(5 \times 10^{-2} \text{cm}) \times (6.5 \times 10^{-5} \text{cm}) \times 3.9 \times 8.84 \times \frac{10^{-14} F}{\text{cm}}}{2.5 \times 10^{-5} \text{cm}} = \textbf{8.6 fF}$$

For the aluminum wires

$$C_{w-s} = 2 \times \frac{(5 \times 10^{-2}\,\text{cm}) \times (6.5 \times 10^{-5}\,\text{cm}) \times 3.9 \times 8.84 \times \frac{10^{-14}\,F}{\text{cm}}}{2.5 \times 10^{-5}\,\text{cm}} = 90\ \text{fF}$$

For the copper wire,

$$C_{w-w} = 55\ \text{fF}$$

8.18 SUMMARY

In this chapter, the process modules of device isolation, contact formation, and interconnection were presented. The simplest isolation techniques involve junction isolation. Various LOCOS-based methods have been widely used, but they suffer from lateral encroachment and incomplete isolation at small junction separations. Trench-based methods have become popular for submicron technologies.

For GaAs technologies, nearly all device isolation is accomplished with semi-insulating substrates.

Conducting islands can be created via proton implantation or mesa etching. Contacts are divided into rectifying and ohmic. For rectifying contacts, the barrier height is a sensitive function of the nature of the metal/semiconductor interface. For ohmic contacts, achieving a low contact resistivity requires a large doping concentration at the metal/semiconductor interface.

Most silicon technologies achieve heavy doping by implantation. Self-aligned silicides (salicides) have been developed to reduce the series resistance of shallow junctions in silicon. Many GaAs technologies use alloyed contacts, sometimes with an implantation, to achieve acceptably low specific contact resistivity's. High performance interconnect requires the use of low resistivity metal on top of low capacitance dielectrics. Aluminum alloys are the most widely used for silicon-based technologies, although copper is now beginning to replace it. Gold is generally used for GaAs technologies. CVD SiO_2 is the most commonly used dielectric, although lower permittivity films such as polyimide are being developed for future applications.

PROBLEMS

1. An advanced metallization process is proposed for high density silicon-based ICs. This process will use several new materials. Identify one advantage and one disadvantage for each new material: (a) CVD tungsten, (b) electroplated copper, and (c) spin-on polyimide.

2. Calculate the specific contact resistance at room temperature for a contact to 10^{17} cm^{-3} n-type GaAs if the metallization has a barrier height of 0.8eV.

3. Show that scaling down to smaller line widths does not affect the RC time constant except for the effects of fringing fields.

4. State the reason why void in a contact via is not acceptable.

5. State a method to reduce aluminum diffusion into silicon substrate.

6. State a method to reduce electro-migration of aluminum.

7. Why silicon oxide film has compression stress at room temperature?

8. A four point probe used to measure the sheet resistance n-type silicon has forcing current of 0.4 mA and measured voltage of 10 mV. Find the sheet resistance of n-type silicon.

REFERENCES

1. J.D. Pummer, M.D. Del, and Peter Griffin, "Silicon VLSI Technology: Fundamentals, Practices, and Modeling", Prentice Hall, (2000).

2. Hong Xiao, "Introduction to Semiconductor Manufacturing Technology," Pearson Prentice Hall, (2001).

3. Debaprasad Das, "VLSI Design," Oxford University Press, (2011).

4. P.B. Ghate, "Electromigration Induced Failures in VLSI Interconnects," *Proc. IEEE 20th Int. Rel. Phys. Symp.*, 292 (1982).

5. J.M. Towner and E.P. van de Ven, "Aluminum Electromigration Under Pulsed D.C. Conditions," *21stAnnu. Proc. Rel. Phys. Symp.*, 36 (1983).

6. J.A. Maiz, "Characterization of Electromigration Under Bidirectional (BDC) and Pulsed Unidirectional Currents," *Proc. 27th Int. Rel. Phys. Symp.*, 220 (1989).

7. S. Vaidya, T.T. Sheng, and A.K. Sinha, "Line Width Dependence of Electromigration in Evaporated Al-0.5%Cu," *Appl. Phys. Lett.* 36:464 (1980).

8. T. Turner and K. Wendel, "The Influence of Stress on Aluminum Conductor Life," *Proc. IEEE Int. Rel. Phys. Symp.*, 142 (1985).

9. H. Kaneko, M. Hasanuma, A. Sawabe, T. Kawanoue, Y. Kohanawa, S. Komatsu, and M. Miyauchi, "A Newly Developed Model for Stress Induced Slit-like Voiding," *Proc. IEEE Int. Rel. Phys. Symp.*, 194 (1990).

10. K. Hinode, N. Owada, T Nishida, and K. Mukai, "Stress-Induced Grain Boundary Fractures in Al-Si Interconnects," *J. Vacuum Sci. Technol.* B 5:518 (1987).

11. S. Mayumi, T. Umemoto, M. Shishino, H. Nanatsue, S. Ueda, and M. Inoue, "The Effect of Cu Addition to Al-Si Interconnects on Stress- Induced Open-Circuit Failures," *Proc. IEEE Int. Rel. Phys. Symp.,* 15 (1987).

12. P. Singer, "Double Aluminum Interconnects," *Semicond. Int.* 16:34 (1993).

13. R.A. Levy and M.L. Green, "Low Pressure Chemical Vapor Deposition of Tungsten and Aluminum for VLSI Applications," *J. Electrochem. Soc.* 134:37C (1987).

14. R.J. Saia, B. Gorowitz, D. Woodruff, and D.M. Brown, "Plasma Etching Methods for the Formation of Planarized Tungsten Plugs Used in Multilevel VLSI Metallization," *J. Electrochem. Soc.* 135:936 (1988).

15. D.C. Thomas, A. Behfar-Rad, G.L. Comeau, M.J. Skvarla, and S.S. Wong, "A Planar Interconnect Technology Utilizing the Selective Deposition of Tungsten—Process Characterization," *IEEE Trans. Electron Dev.* 39:893 (1992).

16. T.E. Clark, P.E. Riley, M. Chang, S.G. Ghanayem, C. Leung, and A. Mak, "Integrated Deposition and Etch back in a Multi-Chamber Single Wafer System," *IEEE VLSI Multilevel Interconnect Conf.,* 478 (1990).

17. K. Suguro, Y. Nakasaki, S. Shima, T. Yoshii, T. Moriya, and H. Tango, "High Aspect Ratio Hole Filling by Tungsten Chemical Vapor Deposition Combined with a Silicon Sidewall and Barrier Metal for Multilevel Metallization," *J. Appl. Phys.* 62:1265 (1987).

18. P.E. Riley, T.E. Clark, E.F. Gleason, and M.M. Garver, "Implementation of Tungsten Metallization in Multilevel Interconnect Technologies," *IEEE Trans. Semicond. Manuf,* 3:150 (1990).

9

Packaging

9.1 INTRODUCTION

Packaging is the second last stage and testing is the last stage of semiconductor device fabrication. In the semiconductor industry it is known as simply packaging and sometimes as semiconductor device assembly. Packaging is also known as encapsulation, because the term packaging generally comprises the steps or the technology of mounting and interconnecting of devices. In evolution days of integrated circuits ceramic flat packs, were used by the military for their reliability and small size. Dual in-line Package (DIP) was the first successful commercial package available first in ceramic and later in plastic. Around 1980s pin numbers of VLSI circuits exceeded the limit for DIP packaging. This limitation leads to pin grid array (PGA).

Surface mount package known as SMD or SMT appeared and became widespread in the 1985s. SMT occupies an area about 40–50% less than a corresponding DIP, and thickness that is 70% less. Then semiconductor industry witness Plastic leaded chip carrier (PLCC) packages. In the late 1995s, plastic quad flat pack (PQFP) and thin small-outline packages (TSOP) became the most common for high pin count devices.

Intel and AMD have transitioned from PGA packages to land grid array (LGA) packages on high performance processors. Ball grid array (BGA) packages were invented in 1970s. 1990 witness Flip-chip Ball Grid Array packages (FCBGA). FCBGA permitted higher pin count than other package types available that time. In this package the die is mounted upside-down (flipped) and connects to the package balls via a package substrate.

FCBGA packages allow an array of input-output signals to be distributed over the entire die. Traces out of the die, through the package, and into the printed circuit board have very different electrical properties, compared to on-chip signals. They require special design techniques and need much more electric power than signals confined to the chip itself.

When multiple dies are stacked in one package, it is called System In Package (SiP). The Multichip Module (MCM) refers to multiple dies combined on a small substrate, often ceramic. Today's majority of integrated circuits are packaged in ceramic or plastic insulation. Metal pins or leads are used to make connections with the outside world.

Packaging refers to the set of unique processes that provide electrical connectivity in integrated circuit. Packaging also provide physical protection or mechanical strength to the chip. For a target system, the various components can be integrated simultaneously into one chip. This is known as System on Chip (SoC). There is another term related to package is System on Package (SoP). SoP refers to multiple chips with specific functionality integrated on a package. The following integrated circuit characteristics affect the packaging process

1. Integration level
2. Wafer thickness
3. Chip dimensions
4. Environmental sensitivity
5. Physical vulnerability
6. Heat generation

9.2 PACKAGE TYPES

A wide variety of packages are available for use in VLSI devices. Packages can be roughly divided into two categories:

1. **Hermetic Ceramic Packages**: Here the chip is kept in an environment free from the external environment. For this a vacuum tight enclosure is utilized. Hermetic packages are usually used for high end applications that allow some cost penalties.
2. **Plastic Packages:** Plastic package are not completely free from the external environment. It is encapsulated with resin materials, generally epoxy based resins are used. They are very cost competitive and their popularity continues due to rapid advances in plastic technology.

The most widely used assembly technologies are the plastics technology and hermetic assembly technology. The plastic assembly can be segmented into several package style both for surface mount

9.2 Package Types

assembly technique and through-holemount assembly techniques. Plastic package styles are available in following styles:

- plastic DIP(PDIP)
- plastic QFP(PQFP)
- single Outlined Package(SOP)
- plastic leadless chip carrier (PLCC)

The general overview of the various package types is illustrated in figure 9.1.

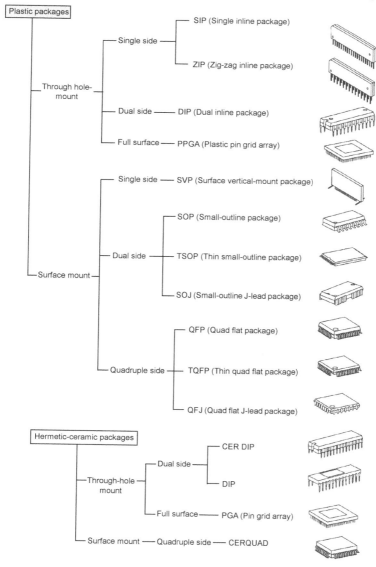

Fig. 9.1 Package Types

For defense, space and industrial applications hermetic assembly technology is basically used. Hermetic technology assemble high reliability integrated circuit. In this technology, the integrated circuit is kept free from external environment by using a vacuum-tight enclosure. Ceramic DIP, BGA and PGA are some of the most widely used packages that are assembled with this technology. Figure 9.2 shows a typical hermetic package with a silicon chip placed in the cavity of a ceramic-based package.

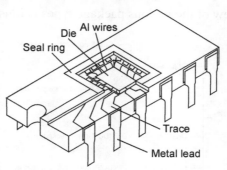

Fig. 9.2 A hermite Package

To assemble low costintegrated circuitin bulk, generally plastic assembly technology is most common. In this technology die is not decoupled from external environment. The die remains in contact with epoxy resin. After long time environment contaminant can penetrate the plastic to reach the integrated circuit. If this happen, it may cause serious reliability issue. However, with today's modern technology, plastic package device are widely acceptable for housing high reliability product. A plastic package structure having silicon die and metal frame, is shown in figure 9.3.

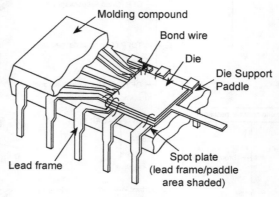

Fig. 9.3 Plastic Package

Most plastic package found their application in electronic devices, memory device usually is low cost. The need of low cost memory device is the driving

9.3 Packaging Design considerations

force behind the cheaper plastic package. A variety of surface mount plastic packages such as SOJ, SOP and thin SOP. TSOP has been developed successfully for industrial use. These plastic packages are manufactured with 2 mm thick body except TSOP. For compact application, TSOP packages are designed with 1 mm thickness. The chip occupancy is growing continuously and the required stringent norms have led to considerable changes in package structures. In the lead-on-chip structure, wire interconnection within the package are made above the die circuitry. This is remarkable achievement. The traditional packages for older-generation devices, are shown in figure 9.2 and figure 9.3. In these packages interconnection is made only outside the die area.

9.3 PACKAGING DESIGN CONSIDERATIONS

Integrated circuit package plays a vital role in the working and efficiency of a component. Package brings electrical signal and voltage supply via wires in and out of the silicon die. it also helps in getting rid of heat generated by the circuit. Package provides mechanical support to strengthen the integrated circuit. It also guards the integrated circuit against extreme environmental conditions like heat and humidity. Also the package left major impact on the power dissipation and efficiency of the integrated circuit like the processor and digital signal processor. This effect is getting more noticeable as technology scaling down is advancing due to reduction of internal signal delays and on-chip parasitic capacitance. Packaging delay are the reason for 50% of the delay in a high-performance computer. These delays are caused by capacitive and inductive parasitic from packing material. With continuous scaling in integrated circuit there is a need for ever more input-output pins. This is because the number of connections is directly proportional to the complexity of the circuitry. E. Rent of IBM, developed an empirical formula known as Rent's rule to demonstrate this relationship. This Rent rule relates the number of input/output pins (P) to the complexity of the circuit. Complexity is measured by the number of components.

$$P = kG^\beta \qquad \ldots (9.1)$$

where k = average number of Inputs/Outputs per component

G = Number of gates. Its value varies from 0.1 and 0.8.

The value of β and k depends on the architecture, circuit organization and application area. This is shown in table 9.1

Chip/System	β	K
Static Memory	0.12	6.00
Gate Array	0.50	1.90

Chip/System	β	K
Microprocessor	0.45	0.80
High Speed Computer Circuit	0.25	82.0
High Speed Computer Chip	0.60	1.45

From this table it is clear that microprocessors possess very different input/output behavior compared to static memory. The observed rate of pin-count increase for IC varies between 9% to 12% per year. Researchers have projected that packages with more than 3,000 pins will be required by the year 2020 end. For all or some reasons, conventional DIP, through-hole mounted packages have been replaced by newer approaches like multichip module, surface-mount and ball grid array techniques. The circuit designer must be aware of the all available options, and their merits and demerits. Having its multi-functionality, a efficient package must own with a large variety of specifications namely the thermal, electrical, thermal, mechanical and cost requirements.

9.4 INTEGRATED CIRCUIT PACKAGE

In general term a package encapsulates the integrated circuit die and splays it out into another device to which one can easily connect. The outer connection on the die is connected through a tiny piece of gold wire to a pin on the package. These are of extreme importance to us, because they are going to connect to wires and other remaining components in the circuit.

Now a day's various types of packages are in use. All package types have unique dimensions, pin-counts and mounting-types.

Mounting Style of Package

How package are mounted to a circuit board is the chief distinguishing package type characteristics. All types of packages fall into two mounting categories:

1. Through-hole package
2. Surface-mount package (SMD or SMT)

It is easy to work with through-hole packages as they are generally bigger in size. They are made to be stuck through one side of a board and soldered to the opposite side.

Unlike through hole package surface-mount packages have small size. They are all made to stuck on one side of a circuit board and soldered to the surface. The pins of a surface mount package are perpendicular to the chip. In some cases, pins are arranged in a matrix on the bottom of the chip. The surface mount packaging is not much suitable for hand-assembly or user friendly. They require special CAD tools to help in the process.

9.4 Integrated Circuit Package

DIP Package

The most common package is DIP (Dual in-line) package. This is now conventional package type. It is type of through-hole IC package. These integrated circuit chips have two parallel rows of pins. The pins in two rows extend perpendicular out of a rectangular, black color, plastic housing. Some DIP packages are shown in figure 9.4.

Fig. 9.4 The 28-pin AT mega 328 micro controller in DIP-package

In a dual in line package two pins are kept 0.1" (2.54 mm) apart. This is a standard spacing. It has proved perfect for fitting into **breadboards**. The pin count decides the overall dimensions of a DIP package. This package may have pins anywhere from 4 to 64.

Apart from being used in breadboards, DIP ICs can also be **soldered into printed circuit board.** They're inserted from one side of the board and soldered on the opposite side of board. Instead of soldering directly to the IC, using a socket for the chip is good idea. With sockets DIP IC can be removed and swapped out easily.

Surface Mount Packages

Surface-mount package are available in many configurations. To build an integrated circuit in surface-mount package, one usually need a custom PCB made for them. They are soldered on a matching pattern of copper.

Small-Outline Package (SOP)

Small-outline integrated circuit (SOIC) packages is a type of SMD/SMT package. It is considered as cousin of the DIP package. These packages are easiest SMD parts that can be hand solder easily. In this package, two pins are usually placed apart nearly 0.05" (1.27 mm).

The other variation available in SMD are

- SSOP – Shrink small-outline package (figure 9.5)
- TSSOP – Thin-shrinksmall-outline package
- TSOP – Thin small-outline package

Fig 9.5: *SSOP package*

Quad Flat Packages

A quad flat package has IC pins in all four directions. QFP ICs contain eight pins per side (32 total pins in all 4 sides) to upwards of seventy-five plus (300+ total in all 4 sides). The two pins on a QFP package are usually kept apart by anywhere from 0.5 mm to 1 mm. There are two more variants available of the standard QFP package. These are

- Thin QFP
- Very thinQFP
- Low-profile QFP

Fig. 9.6 TQFP package

Quad-flat no-leads Package

A **QFN** package is like a QFP package with lags off. The QFN packages have tiny connections. This type of package contains exposed pads on the bottom corner edges of the IC. In some cases they wrap around, and are exposed on both the side and bottom.

Fig. 9.7 QFN package

The three variations Thin QFN, very thin QFN and micro-lead QFN packages are smaller variations of the QFN package. Another package type available, has pins on just two sides. These are

- dual no-lead (DFN) package
- thin-dual no-lead (TDFN) packages.

Now a day many microcontrollers, sensors, and other modern ICs are available in QFP or QFN packages. For example, AT mega 328 microcontroller from Atmel is available in both a TQFP and a QFN-type (MLF) package. While the MPU-6050 comes in a miniscule QFN package.

Ball Grid Arrays Package

Ball grid array (BGA) package are in demand for latest ICs. These are complex little packages. Here little balls of solder are arranged in a 2-D grid on the bottom of the Integrated circuit. In some cases the solder balls are mounted directly to the die.

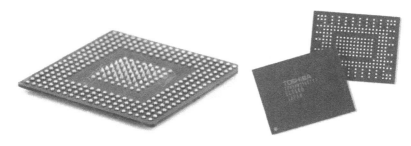

Fig. 9.8 BGA Package

BGA packages are generally used for advanced microprocessors, like Arduino, Raspberry Pi or ARM controllers. To solder a BGA-packaged IC, onto a PCB a automatic pick and place machine is required.

9.5 VLSI ASSEMBLY TECHNOLOGIES

VLSI assembly technologies cover the basic assembly operations in use today for VLSI devices. The assembly flow chart applicable to ceramic or plastic package is shown in figure 9.9.

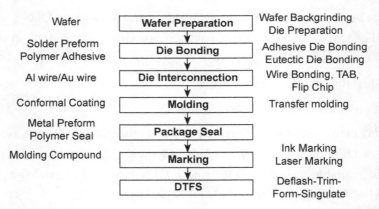

Fig. 9.9 Generic assembly sequence for ceramic and plastic packages.

1. **Wafer Preparation**: In first step wafers undergo cleaning and surface lamination. Prior to assembly, the wafer backside is grinded to the correct wafer thickness. This process is known as Wafer Back grinding. The important grinding wheel parameters out of other are: spindle coolant water temperature, speed, initial and final wafer thickness and flow rate. To get rid of debris wafers are washed continuously during the back grinding process.

2. **Die Interconnection**: In this step a die attach machine is used to pick up the die. Then die machine deposit the wafer on the frame. Advance die machine use the wafer mapping method to pick up only good die. Prior to bonding process, die attach materials like gold, solder wires (lead tin based) or silver epoxy paste are required.

Fig. 9.10 Die interconnections process machine

Either Gold or Aluminum wires are used depending on specification of circuit. The wire is fed through a ceramic capillary. By maintaining good combination of ultrasonic energy and temperature, a good metalized wire bond is formed.

9.5 VLSI assembly Technologies

3. **Die Bonding**: In Die Bonding process the silicon chip is attached with die pad of the supporting structure. For attaching silicon chip adhesives material such as polyimide, epoxy and silver-filled glass are used. While in another die attach method, a eutectic alloy is utilized to attach the die to the cavity or pad. The Au-Si alloy is the most preferable material for this. This method is shown in figure 9.11.

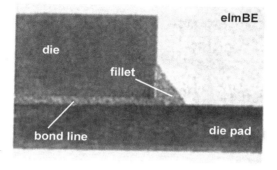

Fig. 9.11 D/A Adhesive as the grainy material

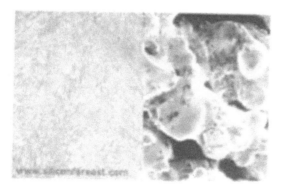

Fig. 9.12 Normal Eutectic die attach and with balling

4. **Wire Bonding**: Wire bonding is performed on ICs after they have been die-bonded to the appropriate piece part. The major issue in establishing a quality and reliable wire bonding process is process control. Two types of process tests are in use today. These are the wire-pull test and the ball-shear test. Wire bond pull tests have been around a long time for evaluating both ball and wedge bonds. The commonly used material for wire is

- Gold
- Aluminum
- Copper

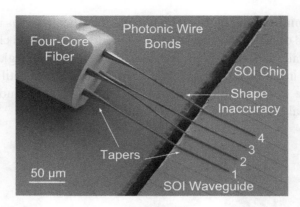

Fig. 9.13 Wire Bonds

The wire bonding may be divided in two categories:

Ball bonding: The extreme end of the wire is melted, to form a gold ball. Then free clean air ball is brought into contact with the bond pad, next adequate amounts of heat, ultrasonic forces and pressure are applied. Then the wire is run to the equivalent finger of the lead frame. This forms a gradual arc between the lead finger and bond pad. Figure 9.14 shows bonding of gold wire to the lead and die.

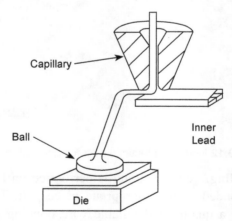

Fig. 9.14 A gold wire bonding (Courtsey - Microchip fabrication)

Wedge bonding: In this bonding style, a clamped wire is carried in contact with the bond pad. Then pressure and controlled ultrasonic energy is applied. Then the wire is run to the equivalent lead finger, and again pressed. Again ultrasonic energy is applied to the wire to form second bond.

Table 9.2 Summary of Intel's Package I/O Lead Electrical Parasitics for Multilayer Packages

Electrical Parameter	Wirebond Package Type			Flip-chip Package Type	
	CPGA	PPGA	H-PBGA	OLGA	FC-PGA
Bondwire/Die bump R (mohms)	126–165	136–188	114–158	2	0.06
Bondwire/Die bump L (nH)	2.3–41	2.5–4.6	2.1–4.1	0.02	0.013
Trace R (mohms/cm)	1200	66	66	590	120
Trace L (nH/cm)	4.32	3.42	3.42	3.07	2.329
Trace C (pF/cm)	2.47	1.53	1.53	1.66	1.707
Trace Z_0 (ohms)	42	47	47	43	38.5
Pin/Land R (mohms)	20	20	0	8	20
Pin/Land L (nH)	4.5	4.5	4.0	0.75	2.9
Plating Trace R (mohms/cm)	1200	66	66	N/A	N/A
Plating Trace L (nH/cm)	4.32	3.42	3.42	N/A	N/A
Plating Trace C (pF/cm)	2.47	1.53	1.53	N/A	N/A
Plating Trace Z_0 (ohms)	42	47	47	N/A	N/A
Trace Length Range (mm)	8.83–26.25	6.60–42.64	4.41-22.24	3.0-18.0	10.0-42.6
Plating Trace Length Range (mm)	1.91-10.50	1.91-16.46	0.930-8.03	N/A	N/A

9.6 YIELD

The manufacturer of integrated circuit is ultimately interested in how many finished chips will be available for sale. A substantial fraction of the dice on a giver wafer will not be functional when they are tested at the wafer probe step at the end of the process. Additional dice will be lost during the die separation and packaging operations, and a number of the packaged devices will fail final testing.

The cost of packaging and testing is substantial and may be the dominant in the manufacturing cost of small die. For a large die with low yield, the manufacturing cost will be dominated by the wafer processing cost. A great deal of time has been spent attempting to model wafer yield associated with IC processes. Wafer yield is related to the complexity of the process and is strongly dependent on the area of the IC die.

Yield is a important parameter in chip fabrication. If one makes thousand chips and if only nine hundred out of the one thousand chips pass all the tests, then the yield will be 90%. Preferably, everyone wants to get 100% yield. But usually, for memory chips, the yield rate is larger than 95% and for processor chips, the yield rate is in range of 60 to 80% assuming that the process and design are reasonable. The following parameters are important to analyze yield performance.

1. **Directivity:** This is quantitative relationship between defect size and the yield loss.
2. **Defect Density:** This explore cause of chip failure also known as root cause analysis, so that the fabrication can take required remedial action.

Defectivity: This parameter is closely related with the yield. It shoes the level of defects generated in a fabrication. The all integrated circuit chips are manufactured in clean rooms with international standards, but even in clean rooms, it is impossible to maintain 100% cleanliness. Even after utmost care, there are some small defects or particles floating around. During fabrication process, some particles settle down on the wafer. The two parameters defect size distribution (DSD) and the number of defects are vital parameters in yield.

In a properly maintained clean room, the defect level will decrease with particle size. The defect density may be defined as number of defects per unit area.

Uniform Defect Densities

One can visualize how die area affects yield by looking at the wafer in figure 9.15. Which has 120 die sites. The dots represent randomly distributed defects that have caused a die to fail testing at the wafer-probe step. In figure 9.15 (a), there are 52 good dice out of the total of 120, giving a yield of 43%. If the die size were twice as large, as in figure 9.15 (b), the yield would be reduced to 22% for this particular wafer.

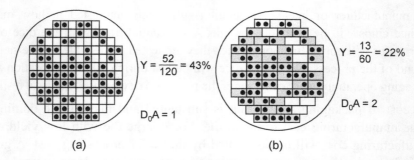

Fig. 9.15 Wafers showing effect of die size on yield. Dots indicates the presence of effective die location (a) For a particular die size yield is 43% (b) if die size is doubled yield becomes 22%

An estimate of the yield of good die can be found from a classical problem in probability theory in which n defects are randomly placed in N die sites. The probability P_k, that a given die site contains exactly k defects is given by the binomial distribution:

$$P_k = \frac{n!}{k!(n-k)!} N^{-n(N-1)^{n-k}} \qquad \ldots (9.2)$$

9.7 SUMMARY

Following the completion of processing, wafers are screened by checking various processing and device parameters using special test sites on the wafer. If the parameters are within proper limits, each die on the wafer is tested for functionality.

Next, the dice are separated from the wafer with a diamond saw or a scribe and break process. Some die loss is caused by damage during the separation process. The remaining good dice are mounted in ceramic or plastic DIPs, LCCs, PGAs, SMDs or BGA package using epoxy or eutectic die-attachment techniques.

Bonding pads of the die are connected to leads on the package using ultrasonic or thermosonic bonding of 15 to 75 µm aluminum or gold wire. Batch fabricated flip chip and TAB interconnection process that permit simultaneous formation of hundreds or even thousands of bonds can also be used.

The final manufacturing cost of an integrated circuit is calculated by the unit of functional parts produced. The overall yield is the ratio of the number of working packaged dice to the original number of dice on the wafer. Yield loss is due to defects on the wafer, processing errors, damage during assembly, and lack of full functionality during final testing. The larger the die size, the lower will be the number of good dice available from a wafer.

PROBLEMS

1. What is packaging? Write short note on package types.
2. What do you understand by DIP package? Explain in brief.
3. Write a detailed note on assembly technologies Explain significance of each stage.
4. What are the Package types in VLSI? Describe packaging design consideration in VLSI.
5. What is Assembly technique? Write the Applications also.
6. In a VLSI/ULSI design, how packaging is evaluated? Elaborate the packaging design consideration.

REFERENCES

1. E. Suhir, "Calculated Thermally Induced Stresses in Adhesively Bonded and Soldered Assemblies", ISHM International Symposium on *Microelectronics*, Atlanta, Georgia, (1986).
2. "Guidelines for Accelerated Reliability Testing of Surface Mount Solder Attachments", IPCSM-785, (1992).

3. M. Born and E. Wolf, "Principles of Optics," 286-300, Pergamon Press, Oxford (1980).

4. Y. Guo, and S. Liu, "Development in Optical Methods for Reliability Analyses in Electronic Packaging Applications," *Experimental/Numerical Mechanics in Electronic Packaging*, 2:10-21 Bellevue. WA, (1997).

5. D. Post, B. Han, and P. Ifju, "High Sensitivity Moire: Experimental Analysis for Mechanics and Materials," Springer-Verlag, Inc., New York (1994).

6. J.W. Stafford, "The Implications of Destructive Wire bond Pull and Ball Bond Shear Testing on Gold Ball-Wedge Wire Bond Reliability," *Semiconductor International*, 82, (1982).

7. W.C. Till and J.T. Luxon, "Integrated Circuits: Materials, Devices and Fabrication", Prentice-Hall, Englewod Cliffs, NJ, (1982).

8. B.T. Murphy, "Cost Size Optima of Monolithic Integrated Circuits," *Proceedings of the IEEE*, 52:1537 (1964).

9. R.B. Seeds, "Yield and Cost Analysis of Bipolar LSI", *IEE IEDM Proceedings*, 12, (1967).

10. The International Technology Roadmap for Semiconductors, The Semiconductor Industry Association (SIA), San Jose, CA:1999, (http:// www.semichips.org).

❑❑❑

10

VLSI Process Integration

10.1 INTRODUCTION

G. W. A. Dummar in 1952 recognized that electronic devices could be made from the single layers of conducting, insulating, amplifying and rectifying material. It was the first description in the integrated circuit development. The first circuit was germanium transistor, resistors and capacitors formed on a wafer.

A modern VLSI fabrication development might be considered to have anywhere from 10^2 to 10^4 steps. Even transistors fabrication themselves is very complex as well, and varies drastically from process to process. Some examples of why fabrication step varies:

1. Will there be only a single type of PMOS transistor and a single type of NMOS transistor in the IC? Very unlikely in a complex modern design. If so, multiple lithography masks and multiple other steps may be necessary to independently pattern and dope the different types of devices.
2. Multigate devices like Graphene FET, G-MOSFET, FinFET might involve a more complex development to manufacture than traditional planar CMOS transistors.
3. In modern process technologies (e.g. 14-32 nm), the pitches of transistor gates, contacts, and other features are much less than the wavelength of light used to pattern them (typically 193 nm immersion lithography).

10.2 FUNDAMENTAL CONSIDERATIONS FOR IC PROCESSING

The process of constructing devices in a single piece of silicon crystal is a process of building successive layers of insulating, conducting and semiconducting material. Each layer is patterned to provide a distinct function and relationship with nearby areas and subsequent layers. The layers are manufactured and patterned by using the technique in the first 9 chapters of this book.

Building Individual Layers

Figure 10.1 shows some of the more important methods to create a layer in or on the silicon crystal. The layers are build by oxidation, implantation, deposition or epitaxial growth of silicon. Each of these layers can be created in two ways:

1. Uniform way
2. Selective way

Uniformly formation is shown in left side of figure and selectively formation is shown in the right side of the figure.

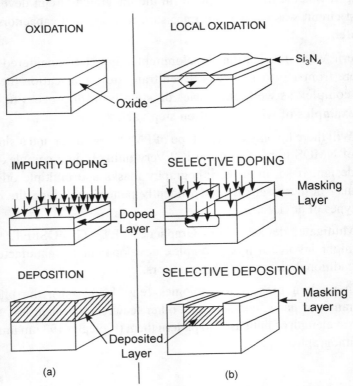

Fig. 10.1 Layers Formation in Silicon (a) uniform method (b) selective method

Photolithography and etching procedures are used to achieve patterning of the layers for selective formation. The various methods for obtaining individual layers have their own usefulness for specific applications. For example, if an insulating layer is required, the oxidation of silicon forms a layer of SiO_2. However the formation of SiO_2 layer consumes silicon and long thermal cycles for thicker layers. Therefore for a thick insulating oxide layer it might require an oxide deposition. The oxide deposited has a poor uniformity as compared to thermally grown oxide, but can be deposited at a lower temperature. At each step in process development, such analysis must be made to find which layer will give the desired results without affecting previous layers. Oxidation and deposition are the most common methods of forming a dielectric layer. The oxidation can be produced selective by depositing and patterning Si_3N_4 before oxidation. This technique is known as local oxidation.

Integrating the Process Steps

In order to build layers requires careful consideration of each layer's relative position to the others. Figure 10.2 shows example of this process of building layers. This schematic illustrates the sequence of forming a simple MOS capacitor.

Figure 10.2(a) shows initial oxidation formation. Then patterned process ia applied on oxide and etched to form the active region of the capacitor.

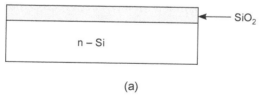

(a)

The active region is formed by ion implanting boron into the silicon exposed by the etching of the oxide as shown in figure 10.2(b)

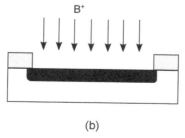

(b)

The subsequent step of growing and patterning a thin oxide is shown in figure 10.2 (c) and figure 10.2 (d) shows depositing and patterning aluminum.

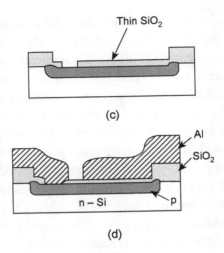

Fig. 10.2 Layers Integration

These steps describes consideration of integrating layers – alignment. The MOS capacitor that results from this process will only work properly if the individual layers are isolated except where an intentional contact is made.

Miniaturizing VLSI Circuits

The minimum feature size on each layer is determined by the ability to reproduce and resolve the feature routinely. This dimension is includes both the minimum dimension and how accurately that feature can be transferred into the silicon during the pattern transfer process. The pattern transfer process will affect the minimum feature size differently. The dry etching causes very little change in the dimensions of the feature etched in the patterned layer. The local oxidation causes the feature to grow as the oxide becomes thicker. Ion implantation involve lateral scattering, increasing the patterned feature size from the mask dimension. Dopants diffusion increases the feature size of a doped region of silicon. All these factors determines the final feature size and must be taken into account when mask dimensions are determined.

10.3 NMOS IC TECHNOLOGY

There are a large diversity of basic fabrication steps used in the fabrication process of modern MOS Integrated circuits. The same fabrication process can be used for the designed of CMOS/BiCMOS devices. The commonly used base material is silicon-on-sapphire (SOS). To avoid the latch up problem i.e. presence of parasitic components like transistors, some variations are introduced in the techniques.

10.3 NMOS IC Technology

An nMOS process fabrication steps may be outlined as follows:

Step 1:

Processing is carried on single crystal silicon of high purity on which required P impurities are introduced as crystal is grown. Such wafers are about 75 to 150 mm in diameter and 0.4 mm thick and they are doped with say boron to impurity concentration of $10^{15}/cm^3$ to $10^{16}/cm^3$.

Step 2 :

A layer of SiO_2 typically 1 µm thick is grown all over the surface of the wafer to protect the surface, acts as a barrier to the dopant during processing, and provide a generally insulating substrate on to which other layers may be deposited and patterned.

Step 3:

The surface is now covered with a layer of photo resist material. This is deposited onto the wafer and spun uniformly of the required thickness.

Step 4:

This photo resist layer deposited in above step is then exposed to ultraviolet light. For this mask are created. Mask defines regions through which diffusion process take place. Mask is divided into two area covered and open to ultraviolet light. The areas that are exposed to UV radiations become polymerized (hardened). While the areas where diffusion is required are shielded by the mask and remain unaffected.

Step 5:

Etching is next step in fabrication. Those areas which remains unaffected in previous step are etched away. The underlying SiO_2 is also etched away to expose the wafer surface in the window defined by the mask.

Step 6:

Now a thin layer of SiO_2 of 0.2 µm is deposited over entire surface again. Then poly silicon is grown on top of this oxide layer. This poly-silicon is deposited by CVD process. This forms gate structure. While fabricating pattern devices, accurate control of resistivity, thickness and impurity concentration is necessary.

Step 7:

Again photoresist coating is applied and mask is defined. Mask allow poly-silicon to be patterned. After this n-type impurities are diffused through mask to define source and drain. Wafer is heated at very high temperature and passed through a chamber that contains desired n-type impurities. This is diffusion process.

Step 8:

Again a thick layer of SiO_2 is deposited over entire surface. Then this layer is masked with photoresist. This is performed to define the area of source, gate and drain where connections (contact cuts) are to be made.

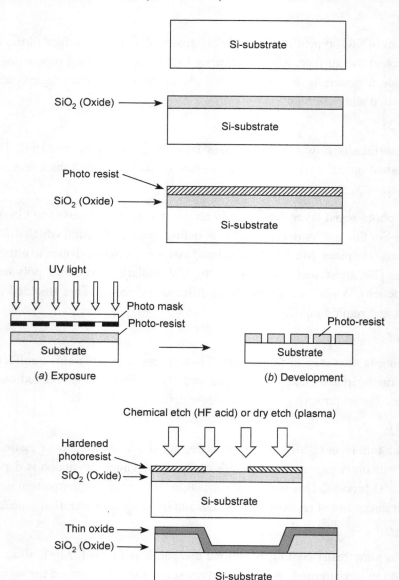

10.3 NMOS IC Technology

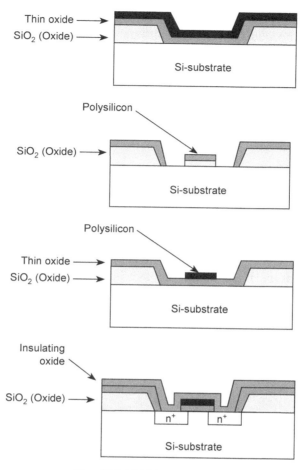

Fig. 10.3 NMOS Fabrication

Step 9:

Now metal generally aluminium is deposited over whole chip. Metal thickness is generally kept at 1 μm. then this aluminum layer is masked and etched to produce the interconnection pattern.

10.4 CMOS IC Technology

In early 1960's Texas started the semiconductor manufacturing process and in 1963, Frank Wanlass got patent for CMOS technology. Since invention CMOS integrated circuits are fabricated by using the semiconductor device integrated fabrication process. These ICs are chief components of almost all electronic and electrical appliances in consumer segment. Most complex and simple electronic circuits are manufactured on a silicon wafer made of semiconductor compounds by utilizing different fabrication steps.

The CMOS technology is used in development of the processors, micro controllers, embedded systems, digital logic circuits and application specific integrated circuits. Its chief advantage is low-power dissipation, full logic swing, high-packing density and very low noise margin. Its most common application is in digital circuitry.

The CMOS IC technology can be fabricated using three different processes. These are:

- N-well process
- P-well process
- Twin tub process

10.4.1 N-Well Process

The n-well fabrication steps are shown in figure 10.4. In the first step mask are used to defines well regions. Then diffusion process is utilized to form n-well at high temperature. Phosphorous implant is used for diffusion process. The depth of well is optimized so that there is no top diffusion breakdown.

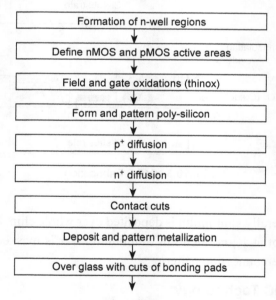

Fig. 10.4 N-well fabrication step

Then devices and diffusion paths are defined, field oxide is grown, poly-silicon is deposit and patterned, diffusions process is carried out, contact cuts are made, and finally metallized.

Figure 10.5 shows an CMOS inverter fabrication. In first step a blank wafer of Si is taken. This wafer is covered with a uniform layer of SiO_2 using oxidation process.

10.3 NMOS IC Technology

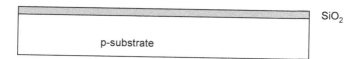

Then the entire SiO$_2$ layer is covered with a layer of photoresist material. At this stage the material is highly insoluble. Now a mask is placed over substrate covered by SiO$_2$ layer and then by photoresist material. Now it is exposed to UV light using the n-well mask. (Photolithography).

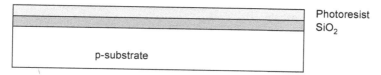

The area exposed to UV lights is removed using organic solvents. This is etching process in fabrication steps.

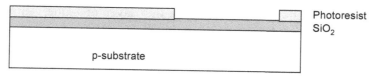

Next etching process is repeated to remove the uncovered oxide using Hydroflouric acid (HF).

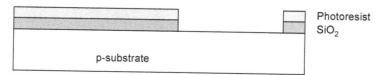

Then using acids remaining photoresist is etched away.

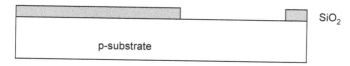

Using diffusion or ion implantation process, n-well is formed within p substrate.

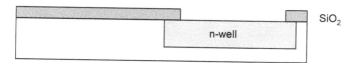

Again using HF acid remaining oxide is etched away. In subsequent steps photolithography process is repeated.

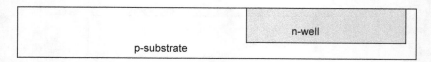

Deposit thin layer of oxide. Use CVD to form poly and dope heavily to increase conductivity

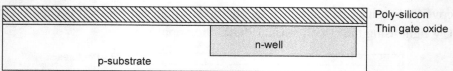

By photolithography process, pattern poly is applied.

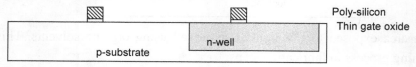

Then the entire surface is covered by a thin layer of oxide. This layer is deposited to produce n diffusion regions.

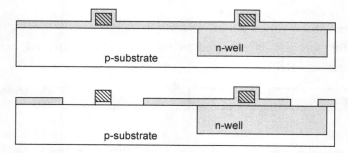

After this diffusion or ion implantation process is utilized to produce n diffusion regions

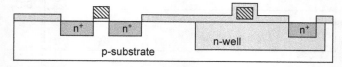

Using etching process the oxide layer is removed to complete patterning step.

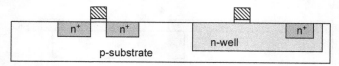

Similar steps used to create p diffusion regions

10.3 NMOS IC Technology

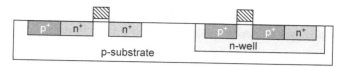

Cover chip with thick field oxide and etch oxide where contact cuts are needed

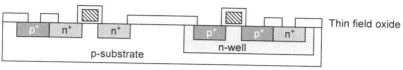

Remove excess metal leaving wires

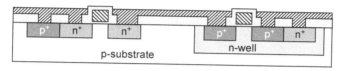

Fig. 10.5 N-Well process

10.4.2 P-Well Process

A brief overview of the fabrication steps may be obtained with reference to figure 10.6, noting that the basic processing steps are of the same nature as those used for nMOS.

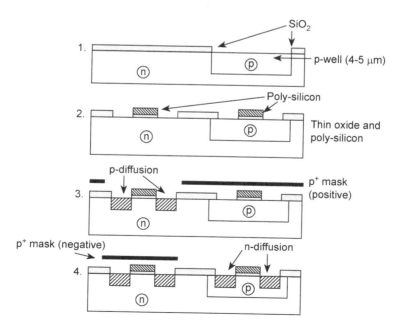

Fig 10.6 Typical steps in CMOS p-well process

p-well process includes suitable masking and diffusion. In this process to lodge n-type devices, a deep p-well is diffused into the n-type substrate.

This fabrication includes two substrste, for this two connections V_{DD} and V_{SS} are required. This is shown in figure 10.7.

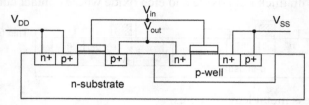

Fig 10.7 View of CMOS p-well inverter component

In most respect, the fabrication step such as masking, patterning, diffusion are similar to pMOS fabrication. The fabrication step are defined in eight step from M1 to M8.

M1: This is mask 1. In this deep p-well diffusion are produced in the region.

M2: Defines the thin oxide regions. In this thin oxide is grown to accommodate wires, n-type transistors and p-type transistors whereas thick oxide is stripped off.

M3 At this step poly-silicon layer is patterned. This layer is deposited after the thin oxide.

M4: All areas where p-type diffusion is to produced, a p+ is used.

M5: In this step n-type diffusion is produced. Negative form of p+ mask is used to obtain this.

M6: Contact cuts are now defined.

M7: This mask defines the metal layer pattern.

M8: Here openings are created for bonding pads. For this a passivation (overglass) layer is now deposited.

10.4.3 Twin Tub Process

The twin tub fabrication process is a logical extension of n and p-well methods.

In this a substrate of n-type material with high resistivity is used. After this n well and p-well regions are formed. This process offer advantage of preserving performance of n-type transistors without degrading p-type transistors. In this doping can be controlled effectively. Latch up can be handled effectively. This process allows separate optimization of p and n-type transistors. Fabrication of inverter using twin tub is shown in figure 10.8

10.5 Bipolar IC Technology

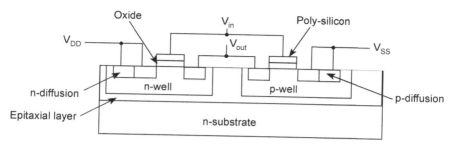

Fig. 10.8 CMOS twin-tub process

10.5 BIPOLAR IC TECHNOLOGY

The base material for a npn transistor is p-type substrate. This substrate is doped at 10^{16} cm^{-3} or less.

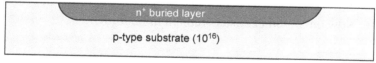

A high dose of n-type dopant is used to form a buried layer. Typically, n-type dopant of phosphorous is used. Either CVD or epitaxial deposition is used, a silicon layer is deposited on the entire surface.

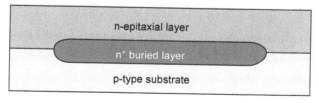

The epitaxial layer so grown forms the collector region of transistor. The highly doped (with low resistivity) buried layer forms an equipotential region. This buried layer is now entirely enclosed in silicon material. Then a very heavy p-type doping is used to form p$^+$ isolation regions.

The junction formed between the n-type epilayer and p+ implant provides electrical isolation.

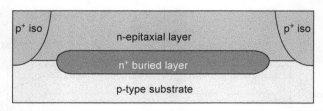

To produce p-type base, the p-type dopant is diffused into the epilayer. Generally boron is used as implant.

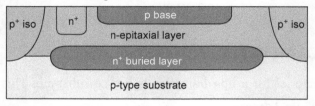

To produce emitter a heavy n-type dopant is diffused through the oxide opening. For transistor operation control of this implant is very important.

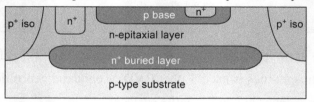

After creating the n+ emitter region in the p-base, the basic BJT structure is complete. Now only metal contacts remain to be form.

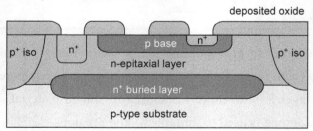

A layer of oxide is grown to isolate metal connections.

After oxide growth using etching, openings are created in the oxide. This oxide opening allow access to the base, emitter and collector regions.

Then metal (generally aluminum) is deposited on entire surface and photoresist is formed on entire surface. A metal mask is used to expose connection area.

Etching is performed to form interconnections between contacts. This forms a complete structure.

10.6 Bi-CMOS Technology

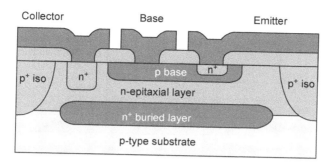

Fig. 10.9 BJT Fabrication

10.6 BI-CMOS TECHNOLOGY

The MOS technology suffers from limited load driving capabilities. The n and p-type transistors have limited current sourcing and sinking capabilities. However we can design super buffers using these MOS transistors but their performance can be enhanced with capabilities of bipolar transistors. Compared with MOS transistors bipolar transistors possess high gain, better noise characteristics and higher frequency. The combination of bipolar and CMOS increases the speed of VLSI circuit.

However, the application of Bi-CMOS in sub-systems such as ALU, ROM, a register-file or a barrel shifter, is not always an effective way of improving speed. This is because most gates in such structures do not have to drive large capacitive loads so that the Bi-CMOS arrangements give no speed advantage.

Bi-CMOS, full potential can be achieved when the whole functional entity and not only single component is considered. A comparison between the characteristics of CMOS and bipolar circuits is set out in Table 10.1 and the differences are self-evident. Bi-CMOS technology goes some way towards combining the virtues of both technologies.

Table 10.1 Comparison of CMOS technology and Bipolar technologies

S No.	Parameter	Bipolar technology	CMOS technology
1	Power dissipation	Very High	Low
2	Input impedance	Low	High
3	Noise Margin	Low	high
4	Package density	Low	High
5	Delay sensitivity to load	Low	High

S No.	Parameter	Bipolar technology	CMOS technology
6	Directional capability	Unidirectional	Bidirectional(source and drain are interchangeable)
7	Voltage swing	High	Low
8	Operating Speed	High	Medium
9	Mask Level	12 to 20	12 to 16
10	Switching Suitability	Not ideal device	Ideal device

Advantages of Bi-CMOS Technology

- Bi-CMOS technology is essentially dynamic to temperature and other process variations parameters offering good economical considerations. This offers very less variability in electrical parameter
- It supports high load current sinking and sourcing as per requirement/specifications..
- As compared to bipolar technique this has low power dissipation
- Bi-CMOS devices are much suitable for Input/Output (I/O) intensive applications, and offers flexible inputs/outputs.
- This technology has improved speed performance as compared to CMOS based technology.
- Latch up invulnerability.
- This has the bidirectional capability that means drain and source can be interchanged.

Applications of Bi-CMOS Technology

The major applications of Bi-CMOS technology are:

- Fit for the intensive input/output applications.
- Initially found applications in RISC processors.
- In microprocessor memory and input/output devices.
- Due to its high input impedance, it is used in sample and hold applications.
- Have applications such as DAC, ADC, mixers, and adders.

10.7 BI-CMOS FABRICATION

The Bi-CMOS device can be manufactured by combining the process of fabrication of BJT and CMOS. The process steps for Bi-CMOS fabrication process are:.

In very first step the p-substrate is covered with a layer of the oxide.

10.7 Bi-CMOS Fabrication

After this a small opening is made on the oxide layer. This opening is used to introduce n-type impurities.

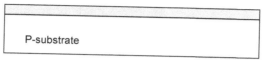

P-substrate with oxide layer

The entire layer is covered again with the oxide layer. Now two openings are made through this oxide layer. Here two n-well are required. From these two openings in oxide layer, the n-type impurities are diffused. This diffused impurities forms n-well.

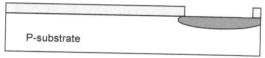

N-type impurities are heavily doped through the opening

To form the three active devices, openings are made via oxide layer. Entire surface is covered with thin oxide and poly-silicon. From this gate terminals pMOS and nMOS are produced.

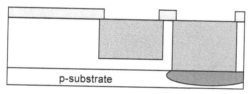

n-type impurities are diffused to form n-wells

The p-impurities are diffused to form the base terminal of bipolar transistor and similar, n-type impurities are diffused to form emitter terminal of BJT, drain and source of nMOS. N-type impurities are diffused into the collector of n-well for contact purpose.

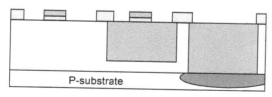

The gate terminals of nMOS and pMOS are
formed with thin oxide and poly-silicon

Then p-type impurities are diffused heavily to produce source and drain of PMOS transistor and to make contact in p-base region.

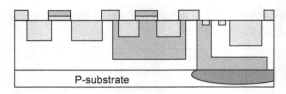

p-type impurities are heavily doped to form
source and drain regions of pMOS

The whole surface is covered again with the thick oxide (SiO$_2$) layer. Through this thick oxide layer the cuts are patterned to form the metal contacts (generally aluminium).

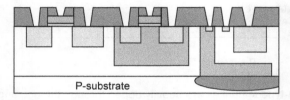

The cuts are patterned to form the metal contacts

The metal contacts are made through the cuts made on oxide layer. In last terminals are named. This is shown in figure 10.10 below.

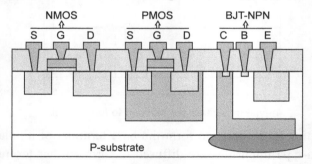

Metal contacts are made through the cuts and terminals
are named

Fig. 10.10 Bi-CMOS Fabrication

10.8 FINFET

Down scaling of traditional MOSFET devices deeper into the micrometer/nanometer side have been threatened by the short channel effects. So there is always thirst for new devices that can overcome short channel effects of conventional MOSFET devices. Some new devices like DG-MOSFET, FinFET and fully depleted SOI MOSFET have promise the possibility of further down scaling of the device. Research has shown that both devices have overcome the problem of short channel effects and latch up.

10.8 FinFET

The basic mode of operation and layout of a FinFET is similar to a traditional field effect transistor. Similar to conventional FET there is one source, one drain and a gate contact to control the current flow.

In opposite to planar MOSFETs the channel is formed as a three-dimensional bar on top of the silicon substrate. This device is called fin because it has a vertically thin channel structure. It resembles a fish's fin. Here gate electrode is wrapped around the channel, due to this several gate electrodes can be produced on each side. This results in enhanced drive current and less leakage current.

FinFET's have broadly been reported to have been fabricated in 2 ways:

- **Gate-first process**: In this fabrication style first the gate stack is patterned/formed and then drain and source are formed.
- **Gate-last process (replacement gate process)**: Here first source and drain regions are formed and then the gate terminal is formed.

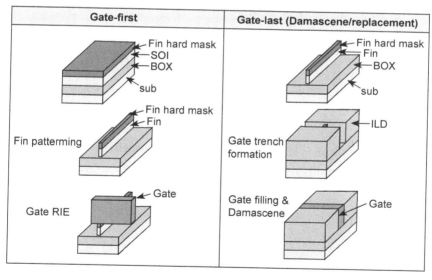

Fig. 10.11 FinFET fabrication processes

Construction of a bulk silicon-based FinFET

1. **Substrate**: A very lightly p-type doped substrate is taken. A hard mask (generally silicon nitride) is used on top of it. Then a patterned resist layer is deposited.

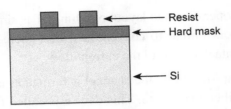

2. **Fin Etch:** A highly anisotropic etching process is used to form the fins. Asin case of Silicon on Insulator, the etch process has to be preciously time based. For the 22 nm process the width of the fins are approximately be 10 to 15 nm, the height is kept at twice.

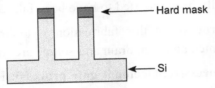

3. **Oxide Deposition**: A thick oxide deposition is made with a high aspect ratio to separate the both fin.

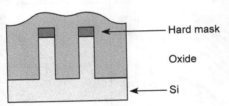

4. **Planarization**: The silicon oxide is planarized by CME (chemical mechanical polishing).The hardmask is utilized as a stop layer.

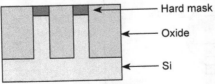

5. **Recess Etch**: To produce the lateralisolation of the fins etch process is used to recess the oxide film.

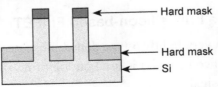

6. **Gate Oxide**: The gate oxide is deposited via thermal oxidation technique to isolate the channel and gate electrode. At this stage the fins are still

connected underneath the oxide. To give complete isolation a high-dose implant at the base of the fin creates a dopant junction.

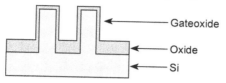

7. **Deposition of the gate**: In last step a highly n+-doped poly silicon layer is deposited on top of the fins. So three gates are wrapped around the channel: one gate is deposited on each side of the fin, and a third gate above the fin.

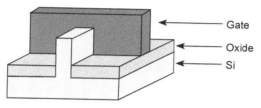

The effect of the top gate can also be inhibited by depositing a silicon nitride layer on top of the channel. Since there is an oxide layer on an SOI wafer, the channels are isolated from each other anyway.

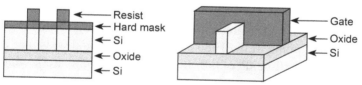

10.9 MONOLITHIC AND HYBRID INTEGRATED CIRCUITS

Based on the method or techniques used in manufacturing them, types of ICs can be divided into three classes:

1. Monolithic ICs
2. Hybrid or multichip ICs

Monolithic IC

A monolithic circuit, literally speaking, means a circuit fabricated from a single stone or a single crystal. The origin of the word 'monolithic' is from the Greek word mono meaning 'single' and lithos meaning 'stone'.

The monolithic integrated circuits are, in fact, fabricated with a single piece of single crystal silicon. The major benefit of integrated circuit of reducing the production cost of electronic semiconductor circuits due to batch production process can be easily understood by a simple example. A standard 10 cm diameter wafer can be divided into approximately 8000 rectangular chips of sides 1 mm.

One IC chip may contain as low as five components to several lakhs components and if 10 such wafers are processed in one batch, we can make 85,000 ICs simultaneously in single go. A lot of chips so fabricated will be faulty due to imperfection in the manufacturing process. Even if the yield (percentage of fault free chips/wafer) is only 20%, it can be seen that 16,000 good chips are produced in a single batch. The production of discrete devices like diodes, transistor, mosfets, or an integrated circuit in general can be achieved by the same technology. The several processes usually take place through a single plane and so, the technology is known as planar technology. .

Hybrid IC

As the name implies, hybrid means, more than one individual types of chips are interconnected. The active components that are contained in this kind of ICs are diffused transistors or diodes. The passive components are the diffused resistors or capacitors on a single chip.

10.10 IC FABRICATION / MANUFACTURING

In the beginning of 1980s, the integrated circuit fabrication was expensive. A typical state of the art high volume manufacturing facility cost over a million dollars. By the beginning of the 1980s, there was deep and widening concern about the economic well-being of the United States. Oil embargoes during the previous decade had initiated two energy crises and caused rampant inflation. The U.S. electronics industry was no exception to the economic downturn, as Japanese companies such as Sony and Panasonic nearly cornered and consumer electronics market.

At that time as of now, the manufacturing of ICs was very expensive. A typical state of the art, high-volume manufacturing facility at that time cost over a 9 million dollars and now costs several billion dollars. In addition, unlike the maker of discrete parts where comparatively little rework is required and a yield larger than 95% on saleable product is frequently realized, the maker of integrated circuits faced unique obstacles. Semiconductor fabrication processes consisted of hundreds of sequential steps, with potential yield loss occurring at every step. So, IC fabrication processes could have yield to lowest value of 20-80%.

Due to continuous increasing costs of fabrication, the challenge before integrated circuit manufacturer was to balance capital investment with automation in the manufacturing process. The target was to use the new developments in computer hardware and software to improve manufacturing methods. So these efforts results in computer integrated fabrication of circuits.

The objective of fabrication of integrated circuits includes higher chip fabrication yield, reduction in product cycle time, maintaining reliable levels of product performance, and improving the reliability of processing equipment. Table 10.2 shows the data of a 1986 study by Toshiba. This study give results on the use of IC-CIM methods in producing 56-K byte DRAM memory circuits.

Table 10.2 1986 Toshiba study result

Productivity Metric	With CIM	Without CIM
Turnaround time	0.58	1.0
Integrated unit output	1.50	1.0
Average equipment uptime	1.32	1.0
Direct labor hours	0.75	1.0

To successfully manufacture VLSI circuit, the process step must be carried out in an environment that is meticulously controlled with respect to cleanliness, temperature humidity, and orderliness. Fabrication monitoring and control are other important areas.

10.11 FABRICATION FACILITIES

In 1965 the chip manufacturing factories were dirty by today's standards and wafer cleaning procedures were orally understood. The chips were manufactured even in those days but they were very small and contained very few components by today's standards. Defects on a chip tens to reduce yields exponentially as chip size increases, small chip can be manufactured even in quite dirty environment.

The semiconductor devices are fabricated by introducing dopants, often at concentrations of parts per billion and by depositing and patterning thin films on the wafer surface, often with a thickness control of a few nm. Such process are manufacturable only if stray contaminants can be held to levels below those that affect devices characteristics on chip yield. Modern IC manufacturing units employ clean rooms to control unwanted impurities. Clean room is implemented by building the chips in a clean dust free environment. The air is highly filtered. Apparatus are designed to minimize particle production. Ultra pure chemicals and gases are used in wafer processing.

The numerous developments have been made in shrinking device geometry and in improving manufacturing so that larger chips can be economically built.

This development requires that defect control associated in the manufacturing process also improve. Lets have a look at the SIA data summarized in table 10.3

Table 10.3 Implication of Semiconductor Industry Growth on defect size, density and contamination level

2015	2012	2009	2006	2003	1999	Year of DRAM shipment
18 nm	25 nm	35 nm	50 nm	65 nm	90 nm	Critical defect size
0.05	0.015	0.03	0.06	0.14	0.29	Starting wafer total LLS (cm^{-2})
0.001	0.001	0.003	0.006	0.014	0.03	DRAM GOI defect density (cm^{-2})
0.01	0.03	0.04	0.05	0.08	0.15	Logic GOI defect density (cm^{-2})
< 10^9	< 10^9	< 10^9	$1*10^9$	$2*10^9$	$4*10^9$	Critical metals on wafer surface after cleaning (cm^{-2})
>=450	>=450	>=450	>=325	>=325	>=325	Starting Material Recombination Lifetime (μsec)
Under $1*10^{10}$	Under $1*10^{10}$	Under $1*10^{10}$	Under $1*10^{10}$	Under $1*10^{10}$	$1*10^{10}$	Standard Wafer total bulk Fe (cm^{-2})

The critical particle size is on the order of half of the minimum feature size. Particles larger than this size have a high probability of causing a manufacturing defect.

It is obvious that great care must be taken in making sure that the factories in which chips are manufactured are as clean as possible. Even with a ultra clean environment, and even with procedure with clean wafers thoroughly and often, it is not realistic to expect that all impurities can be kept out of silicon wafers. There is simply too much processing and handling of the wafers during IC fabrication.

The manufacturing units producing chips must be clean facilities. Particles that might deposit on a silicon wafer and cause a defect may originate from many sources including people, machines chemicals and process gases. Such particles may be airborne or may be suspended in liquids or gases. It is common to characterize the cleanliness of air in IC facilities by the designation "class 10 or class 100". Figure 10.12 illustrates the meaning of these terms.

10.11 Fabrication Facilities

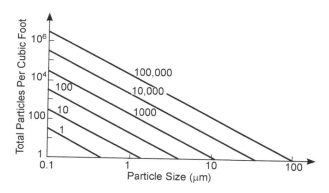

Fig. 10.12 Particle size distribution curve for various classes of clean room. The vertical axis is the total number of particles larger than a given particle size.

Class 10 simply means that in each cubic foot of air in the factory. There are less than 10 total particles greater than 0.5 μm in size. A typical class room of university is about class 100000 while room air in state of the art manufacturing facilities today is typically class 1 in critical areas. This level of cleanliness is obtained through a combination of air filtration and circulation, clean room design and through careful elimination of particular sources.

Particles in the air in a manufacturing plant generally come from several main sources. These includes the people who work in the plant, machines that operate in the plant, and supplies that are brought into the plant. Many studies have been done to identify particle source and the relative importance of various sources. For example people typically emit several hundred particle per minute from each cm^2 of surface area. The actual rate is different for clothing versus skin versus hair but net result is that a typical human emits 5-10 million particles per minute.

Most modern IC manufacturing plant make use of robots for wafer handling in an effort to minimize human handling and therefore particle contamination.

The very first step in introducing particles is to minimize these sources. People in the plant wear "bunny suits" which cover their bodies and clothing and which lock particle emissions from these sources. Often face masks and individual air filters are worn to prevent exhaling particles into the room air. Air showers at the entrance to the clean room blow loose particles off people before they enter and clean room protocols are enforced to minimize particle generation. Machines that handles the wafers in the plant are specifically designed to minimize particle generation and materials are chosen for use inside the plant which minimize particle emission.

The source of particles can never be completely eliminated, constant air filtration is used to remove particles as they are generated. This is accomplished by recirculating the air through High Efficiency Particulate Air (HEPA) filters. These filters are composed of thin porous sheets of ultrafine glass fibers(< 0.5 µm diameter). Room air is forced through the filters with a velocity of about 50 cm/sec. Large particles are trapped by the filters; small particles impact the fibers as they pass through the filter and stick to these fibers primarily through electrostatic forces. The net result is HEPA filters are 99.98% efficient at removing particles from the air.

Most IC manufacturing facility produce their own clean water on site, starting with water from the local water supply. This water is filtered to remove dissolved particles and organics. Dissolved ionic species are removed by ion exchange or reverse osmosis. The result is high purity water that is used in large quantities in the plant.

Modern Chip manufacturing plant are designed to continuously recirculate the room air through HEPA filters to maintain a class 10 or class 1 environment. A typical clean room is shown in figure 10.13.

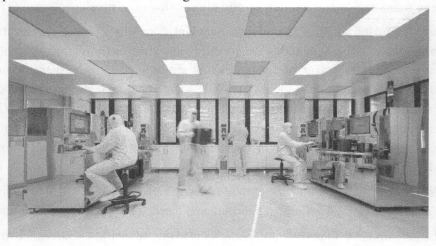

Fig. 10.13 Typical modern cleanroom for IC fabrication.
(Photo courtesy of grapheme manchester.ac.uk)

All mechanical support equipment is located beneath the clean room to minimize contamination to these machines. The HEPA filters are located in the ceiling of clean room the fan that recirculate the air are normally placed above HEPA filter. Inside cleanroom, finger walls or chases provide a path for air return as well as to bring in electric power, distilled water and gases. The scientist and engineers wear "bunny suits" to minimize particle emission.

Process Flow and Key Measurement Points

When we monitor a physical system, we observe that system's behavior. On the basis of these observations, we take appropriate actions to influence that behavior in order to guide the system to some desirable satte. Semiconductor manufacturing systems consist of a series of sequential process steps in which layers of materials are deposited on substrate, doped with impurities, and patterned using photolithography (chapter 4) to produce sophisiticated integrated circuits and devices.

As an example of such an system, figure 10.14 depicts a typical CMOS process flow. Inserted into this flow diagram in various places are symbols denoting key measurement points. Clearly, CMOS technology involves many unit processes with high complexity and tight tolerances. This necessitates frequent and through inline process monitoring to assure high quality final products.

The measurement required may characterize physical parameters such as film thickness, uniformity, and feature dimensions; or electrical parameters, such as resistance and capacitance. These measurements may be performed directly on product wafers, either directly or using test structures, or alternatively, on non functional monitor wafers (or dummy wafers). In addition to these, some measurements are actually performed " in situ" or during a fabrication step. When a process sequence is complete, the product wafer is diced, packaged, and subjected to final electrical and reliability testing.

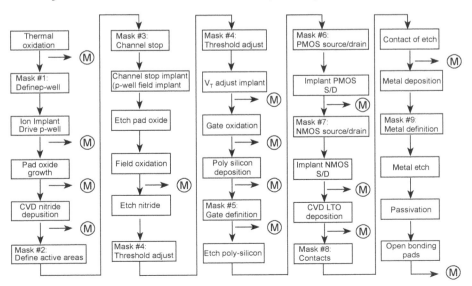

Fig. 10.14 CMOS process flow showing key measurement points

10.12 SUMMARY

This chapter considered processing technologies for passive components, active devices and ICs. Four major IC technologies based on the bipolar transistor, the MOS-FET, CMOS, Bi-CMOS and FinFET were discussed in detail. It appears that the FinFET will be the dominant technology at least until 2025 because of its superior performance compared with the peer component. For 100 nm CMOS technology, a good candidate is the combination of an SOI substrate with interconnections using Cu and low-k materials.

PROBLEMS

1. With a neat sketch explain BiCMS fabrication.
2. Describe various steps in fabrication of CMOS.
3. Discuss N-well process for CMOS fabrication.
4. Discuss advantage of CMOS over bipolar devices.
5. An n well process has thin oxide, n-well and n-plus masl layers, in addition to the other regular layers. Draw the mask combinations to obtain an n transistor, a p-transistor contact, a V_{DD} contact and a V_{SS} contact.

REFERENCES

1. S. Wolf, "Silicon Processing for the VLSI Era, Vol 3," *The Submicron MOSFET,* Lattice Press, Sunset Beach, CA, (1995).
2. S.M. Sze, "Physics of Semiconductor Devices," Wiley, New York, (1981).
3. E.H. Nicollian and J.R. Brews, "*Metal Oxide Semiconductor Physics and Technology,*" Wiley, New York, (1982).
4. Y.P. Tsividis, "Operation and Modeling of the MOS Transistor," McGraw-Hill, New York, (1987).
5. F.M. Wanlass and C.T. Sah, "Nanowatt Logic Using Field-Effect Metal-Oxide-Semiconductor Triodes," *IEEE Int. Solid-State Circuits Conf.,* (1963).
6. J.Y. Chen, "CMOS Devices and Technology for VLSI," Prentice-Hall, Englewood Cliffs, NJ, (1989).
7. R. Chwang and K. Yu, "CMOS—An n-Well Bulk CMOS Technology for VLSI," *VLSI Design,* 42 (1981).
8. L.C. Parrillo, L.K. Wang, R.D. Swenumson, R.L. Field, R.C. Melin, and R.A. Levy, "Twin-Tub CMOS II," *IEDM Tech. Dig.* 706 (1982).

9. R.H. Dennard, F.H. Gaensslen, H.N. Yu, V.L. Rideout, E. Barsous and A. R. LeBlanc, "Design of Ion-Implanted MOSFETs with Very Small Physical Dimensions," *IEEE J. Solid-State Circuits* SC, 9:256 (1974).

10. Y. El Maney, "MOS Device and Technology Constraints in VLSI," *IEEE Trans. Electron* Dev.ED, 29:567 (1982).

11. J.R. Brews, W. Fichtner, E.H. Nicollian, and S.M. Sze, "Generalized Guide for MOSFET Miniaturization," *IEEE Electron Devices Lett.* EDL 1:2 (1980).

12. M.H. White, F. Van de Wiele, and J.P. Lambot, "High-Accuracy Models for Computer-AidedDesign," *IEEE Trans. Electron Dev.* ED, 27:899 (1980).

13. P.I. Suciu and R.I. Johnston, "Experimental Derivation of the Source and Drain Resistance of MOS Transistors," *IEEE Trans. Electron Dev.* ED, 27:1846 (1980).

14. M.C. Jeng, J.E. Chung, P.K. Ko, and C. Hu, "The Effects of Source/ Drain Resistance on Deep Submicrometer Device Performance," *IEEE Trans. Electron Dev.* 37:2408 (1990).

15. C.Y. Lu, J.M.J. Sung, R.Liu, N.S. Tsai, R. Singh, S.J. Hillenius, and G. C. Kirsch, "Process limitation and Device Design Trade-offs of SelfAligned $TiSi_2$ Junction Formation in Submicrometer CMOS Devices," *IEEE Trans. Electron Dev.* 38:246 (1991).

16. B. Davari, W.H. Chang, K.E. Petrillo, C.Y. Wong, D. Moy, Y. Taur, M.W. Wordeman, J.Y.C. Sun, C.C.H. Hsu, and M.R. Polcari, "A High Performance 0.25 mm CMOS Technology: II—Technology," *IEEE Trans. Electron Dev.* 39:967 (1992).

17. S. Nygren and F. d'Heurle, "Morphological Instabilities in Bilayers Incorporating Polycrystalline Silicon," *Solid State Phenom.* 23&24:81 (1992).

18. A. Ohsaki, J. Komori, T. Katayama, M. Shimizu, T. Okamoto, H. Kotani, and S. Nagao, "Thermally Stable $TiSi_2$ Thin Films by Modification in Interface and Surface Structures," *Ext. Absstr. 21st SSDM,* 13 (1989).

19. C.Y. Ting, F.M. d'Heurle, S.S. Iyer, and P.M. Fryer, "High Temperature Process Limitationson $TiSi_2$," *J. Electrochem. Soc.* 133:2621 (1986).

20. H. Sumi, T. Nishihara, Y. Sugano, H. Masuya, and M. Takasu, "New Silicidation Technology by SITOX (Silicidation Through Oxide) and Its Impact on Sub-Half-Micron MOS Devices," *Proc. IEDM,* 249 (1990).

21. F.C. Shone, K.C. Saraswat and J.D. Plummer, "Formation of a 0.1 m n/p and p/n Junction by Doped Silicide Technology," *IEDM Tech. Dig.,* 407 (1985).

22. R. Liu, D.S. Williams, and W.T. Lynch, "A Study of the Leakage Mechanisms of Silicided n+/p Junctions," *J. Appl. Phys.* 63:1990 (1988).

23. M.A. Alperin, T.C. Holloway, R.A. Haken, C.D. Gosmeyer, R.V. Karnaugh, and W.D. Parmantie, "Development of the Self-Aligned Titanium Silicide Process for VLSI Applications," *IEEE J. Solid-State Circuits* SC, 20:61 (1985).

24. R. Pantel, D. Levy, D. Nicholas, and J.P. Ponpon, "Oxygen Behavior During Titanium Silicide Formation by Rapid Thermal Annealing," *J. Appl. Phys.* 62:4319 (1987).

25. D.B. Scott, W.R. Hunter, and H. Shichijo, "A Transmission Line Model for Silicided Diffusions: Impact on the Performance of VLSI Circuits," *IEEE Trans. Electron Dev.* ED, 29:651 (1982).

26. P. Liu, T.C. Hsiao, and J.C.S. Woo, "A Low Thermal Budget Self-Aligned Ti Silicide Technology Using Germanium Implantation for Thin-Film SOI MOSFETs," *IEEE Trans. Electron. Dev.* 45(6):1280 (1998).

27. J.A. Kittl and Q.Z. Hong, "Self-aligned Ti and Co Silicides for High Performance sub-0.18_mCMOS Technologies," *Thin Solid Films* 320:110 (1998).

Appendix A
Properties of Ge and Si at 300 K

Properties	Ge	Si
Atoms/cm^3	4.42×10^{22}	5.0×10^{22}
Atomic weight	72.60	28.09
Breakdown field (V/cm)	$\sim 10^5$	$\sim 3 \times 10^5$
Crystal structure	Diamond	Diamond
Density (g/cm^3)	5.3267	2.328
Dielectric constant	16.0	11.9
Effective density of states in conduction band, N_C (cm^{-3})	1.04×10^{19}	2.8×10^{19}
Effective density of states in valence band, N_V (cm^{-3})	6.0×10^{18}	1.04×10^{19}
Effective Mass, m^*/m_0 Electrons	$m_l^* = 1.64$ $m_t^* = 0.082$	$m_l^* = 0.98$ $m_t^* = 0.19$
Effective Mass, m^*/m_0 Holes	$m_{lh}^* = 0.044$ $m_{hh}^* = 0.28$	$m_{lh}^* = 0.16$ $m_{hh}^* = 0.49$
Electron affinity, χ(V)	4.0	4.05
Energy gap (eV) at 300K	0.66	1.12
Intrinsic carrier concentration (cm^{-3})	2.4×10^{13}	1.45×10^{10}
Intrinsic Debye length (mm)	0.68	24
Intrinsic resistivity (Ω-cm)	47	2.3×10^5
Lattice constant (Å)	5.64613	5.43095
Linear coefficient of thermal expansion, $\Delta L/LDT$ (°C^{-1})	5.8×10^{-6}	2.6×10^{-6}
Melting point (°C)	937	1415
Minority carrier lifetime (s)	10^{-3}	2.5×10^{-3}
Mobility (drift) (cm^2/V-s)	3900 1900	1500 450
Optical-phonon energy (eV)	0.037	0.063
Phonon mean free path l_0 (Å)	105	76 (electron) 55 (hole)
Specific heat (J/g-°C)	0.31	0.7
Thermal conductivity at 300 K (W/cm-°C)	0.6	1.5
Thermal diffusivity (cm^2/s)	0.36	0.9
Vapor pressure (Pa)	1 at 1330°C 10^{-6} at 760°C	1 at 1650°C 10^{-6} at 900°C

Appendix B
List of Symbols

Symbol	Unit	Description
m^*	Kg	Effective mass
m_0	Kg	Electron rest mass
L	cm or μm	Length
kT	ev	Thermal Energy
k	J/K	Boltzman constant
J	A/cm^2	Current density
I	A	Current
hv	eV	Photon energy
h	J-s	Planck's constant
f	Hz	Frequency
ε_m	V/cm	Maximum field
ε	V/cm	Electric field
E_g	eV	Energy bandgap
E_F	eV	Fermi energy level
E	eV	Energy
D	cm^2/s	Diffusion coefficient
C	F	Capacitance
c	cm/s	Speed of light in vacuum
B	Wb/m^2	Magnetic induction
a	Å	Lattice constant
φ	V	Barrier height or imref
ρ	Ω-cm	resistivity
μ_p	cm^2/V-s	Hole mobility
μ_n	cm^2/V-s	Electron mobility
μ_0	H/cm	Permeability in vacuum
τ	s	Lifetime or decay time

Appendix C

Useful Physical Constants

Quantity	Value	Symbol
Wavelength of 1 eV quantum	1.23977 μm	λ
Thermal voltage at 300 K	0.0259 V	kT/q
Thousandth of an inch	25.4 μm	mil
Torr	1 mm Hg	
Standard atmosphere	1.01325×10^5 N/m^2	
Speed of light in vacuum	2.99792×10^{10} cm/s	c
Photon rest mass	1.67264×10^{-27} kg	M_p
Reduced Planck constant	1.05458×10^{-34} J-s (h/2π)	
Room temperature value of kT	0.0259 eV	
Micron	10^{-4} cm	μm
Gas constant	1.98719 cal-mol^{-1}K^{-1}	R
Gram-mole	6.023×10^{23} molecules	
Electron volt	1.60218×10^{-19}	eV
Electron rest mass	0.91095×10^{-30} kg	m_0
Elementary charge	1.60218×10^{-19} C	q
Calorie	4.184 J	
Bohr radius	0.52917 Å	a_B
Boltzman constant	1.38066×10^{-23} J/K	k
Avogadro constant	6.02204×10^{23} mol^{-1}	N_{AVO}
Atmosphere	760 mmHg	
Angstrom unit	1 Å = 10^{-1} nm	Å

Appendix D

Periodic table of the elements and element electronic mass

B = Solids   Hg = Liquids   Kr = Gases   Pm = Not found in nature

1	2	3	4	5	6	7	8	9	10	11	12	13	14	15	16	17	18
1 H 1.00794																	2 He 4.002602
3 Li 6.941	4 Be 9.012182											5 B 10.811	6 C 12.0107	7 N 14.00674	8 O 15.9994	9 F 18.9984032	10 Ne 20.1797
11 Na 22.989770	12 Mg 24.3050											13 Al 26.581538	14 Si 28.0855	15 P 30.973761	16 S 32.066	17 Cl 35.4527	18 Ar 39.948
19 K 39.0983	20 Ca 40.078	21 Sc 44.955910	22 Ti 47.867	23 V 50.9415	24 Cr 51.9961	25 Mn 54.938049	26 Fe 55.845	27 Co 58.933200	28 Ni 58.6534	29 Cu 63.545	30 Zn 65.39	31 Ga 69.723	32 Gc 72.61	33 As 74.92160	34 Se 78.96	35 Br 79.504	36 Kr 83.80
37 Rb 85.4678	38 Sr 87.62	39 Y 88.90585	40 Zr 91.224	41 Nb 92.90638	42 Mo 95.94	43 Tc (98)	44 Ru 101.07	45 Rh 102.90550	46 Pd 106.42	47 Ag 196.56655	48 Cd 112.411	49 In 114.818	50 Sn 118.710	51 Sb 121.760	52 Te 127.60	53 I 126.90447	54 Xe 131.29
55 Cs 132.90545	56 Ba 137.327	71 Lu 174.967	72 Hf 178.49	73 Ta 180.94.79	74 W 183.84	75 Re 186.207	76 Os 190.23	77 Ir 192.217	78 Pt 195.078	79 Au 196.56655	80 Hg 200.59	81 Tl 204.3833	82 Pb 207.2	83 Bi 208.58038	84 Po (209)	85 At (210)	86 Rn (222)
87 Fr (223)	88 Ra (226)	103 Lr (262)	104 Rf (261)	105 Db (262)	106 Sg (263)	107 Bh (262)	108 Hs (265)	109 Mt (266)	110 Ds (269)	111 Rg (272)	112 Cn (277)	113 Unit (277)	114 Uuq (277)	115 Uup (277)	116 Uuh (277)		118 Uuo (277)

57 La 138.9055	58 Ce 140.116	59 Pr 140.50765	60 Nd 144.24	61 Pm (145)	62 Sm 150.36	63 Eu 151.964	64 Gd 157.25	65 Tb 158.92534	66 Dy 162.50	67 Ho 164.93032	68 Er 167.26	69 Tm 168.93421	70 Yb 173.04
89 Ac 232.0381	90 Th 232.0381	91 Pa 231.035888	92 U 238.0289	93 Np (237)	94 Pu (244)	95 Am (243)	96 Cm (247)	97 Bk (247)	98 Cf (251)	99 Es (252)	100 Fm (257)	101 Md (258)	102 No (259)

Appendix E
Some Properties of the Error Function

$$\operatorname{erf} u = \frac{2}{\sqrt{\pi}} \int e^{-z^2} dz = \frac{2}{\sqrt{\pi}} \left(u - \frac{u^3}{3 \times 1!} + \frac{u^5}{5 \times 2!} - \dots \right)$$

Therefore

$$\operatorname{erf}(-u) = -\operatorname{erf} u$$

$$\operatorname{erfc} u = 1 - \operatorname{erf} u = \frac{2}{\sqrt{\pi}} \int_u^\infty e^{-z^2} dz$$

$$\operatorname{erf} u \approx \frac{2u}{\sqrt{\pi}} \quad \text{for } u \ll 1$$

$$\operatorname{erf} u \approx \frac{1}{\sqrt{\pi}} \frac{e^{-u^2}}{u} \quad \text{for } u \gg 1$$

$$\operatorname{erf}(\infty) = 1, \operatorname{erf}(0) = 0$$

$$\operatorname{erf}(0) = 1, \operatorname{erfc}(\infty) = 0$$

$$\frac{d \operatorname{erf} u}{du} = \frac{2}{\sqrt{\pi}} e^{-u^2}$$

$$\int_0^u \operatorname{erfc} z \, dz = u \operatorname{erfc} u \frac{1}{\sqrt{\pi}} \left(1 - e^{-u^2} \right)$$

$$\int_0^\infty \operatorname{erfc} z \, dz = \frac{1}{\sqrt{\pi}}$$

$$\int_0^\infty e^{-u^2} du = \frac{\sqrt{\pi}}{2}, \int_0^u e^{-z^2} dz = \frac{\sqrt{\pi}}{2} \operatorname{erf} u$$

w	erf(w)	w	erf(w)	w	erf(w)	w	erf(w)
0.00	0.000 000	0.44	0.466 225	0.88	0.786 687	1.32	0.938 065
0.01	0.011 283	0.45	0.475 482	0.89	0.719 843	1.33	0.940 015
0.02	0.022 565	0.46	0.484 655	0.90	0.796 908	1.34	0.941 914
0.03	0.033 841	0.47	0.493 745	0.91	0.801 883	1.35	0.943 762
0.04	0.045 111	0.48	0.502 750	0.92	0.806 768	1.36	0.945 561
0.05	0.056 372	0.49	0.511 668	0.93	0.811 564	1.37	0.947 312
0.06	0.067 622	0.50	0.520 500	0.94	0.816 271	1.38	0.949 016
0.07	0.078 858	0.51	0.529 244	0.95	0.820 891	1.39	0.950 673
0.08	0.090 078	0.52	0.537 899	0.96	0.825 424	1.40	0.952 285
0.09	0.101 281	0.53	0.546 464	0.97	0.829 870	1.41	0.953 852
0.10	0.112 463	0.54	0.554 939	0.98	0.834 232	1.42	0.955 376
0.11	0.123 623	0.55	0.563 323	0.99	0.838 508	1.43	0.956 857
0.12	0.134 758	0.56	0.571 616	1.00	0.842 701	1.44	0.958 297
0.13	0.145 867	0.57	0.579 816	1.01	0.846 810	1.45	0.959 695
0.14	0.156 947	0.58	0.587 923	1.02	0.850 838	1.46	0.961 054
0.15	0.167 996	0.59	0.595 936	1.03	0.854 784	1.47	0.962 373
0.16	0.179 012	0.60	0.603 856	1.04	0.858 650	1.48	0.963 654
0.17	0.189 992	0.61	0.611 681	1.05	0.862 436	1.49	0.964 898
0.18	0.200 936	0.62	0.619 411	1.06	0.866 144	1.50	0.966 105
0.19	0.211 840	0.63	0.627 046	1.07	0.869 773	1.51	0.967 277
0.20	0.222 703	0.64	0.634 586	1.08	0.873 326	1.52	0.968 413
0.21	0.233 522	0.65	0.642 029	1.09	0.876 803	1.53	0.969 516
0.22	0.244 296	0.66	0.649 377	1.10	0.880 205	1.54	0.970 586
0.23	0.255 023	0.67	0.656 628	1.11	0.883 533	1.55	0.971 623
0.24	0.265 700	0.68	0.663 782	1.12	0.886 788	1.56	0.972 628
0.25	0.276 326	0.69	0.670 840	1.13	0.889 971	1.57	0.973 603
0.26	0.286 900	0.70	0.677 801	1.14	0.893 082	1.58	0.974 547
0.27	0.297 418	0.71	0.684 666	1.15	0.896 124	1.59	0.975 462
0.28	0.307 880	0.72	0.691 433	1.16	0.899 096	1.60	0.976 348
0.29	0.318 283	0.73	0.698 104	1.17	0.902 000	1.61	0.977 207
0.30	0.328 627	0.74	0.704 678	1.18	0.904 837	1.62	0.978 038
0.31	0.338 908	0.75	0.711 156	1.19	0.907 608	1.63	0.978 843
0.32	0.349 126	0.76	0.717 537	1.20	0.910 314	1.64	0.979 622
0.33	0.359 279	0.77	0.723 822	1.21	0.912 956	1.65	0.980 376
0.34	0.369 365	0.78	0.730 010	1.22	0.915 534	1.66	0.981 105
0.35	0.379 382	0.79	0.736 103	1.23	0.918 050	1.67	0.981 810
0.36	0.389 330	0.80	0.742 101	1.24	0.920 505	1.68	0.982 493
0.37	0.399 206	0.81	0.748 003	1.25	0.922 900	1.69	0.983 153
0.38	0.409 009	0.82	0.753 811	1.26	0.925 236	1.70	0.983 790
0.39	0.418 739	0.83	0.759 524	1.27	0.927 514	1.71	0.984 407
0.40	0.428 392	0.84	0.765 143	1.28	0.929 734	1.72	0.985 003
0.41	0.437 969	0.85	0.770 668	1.29	0.931 899	1.73	0.985 578
0.42	0.447 468	0.86	0.776 110	1.30	0.934 008	1.74	0.986 135
0.43	0.456 887	0.87	0.781 440	1.31	0.936 063	1.75	0.986 672

w	erf(w)	w	erf(w)	w	erf(w)	w	erf(w)
1.76	0.987 190	2.22	0.998 308	2.67	0.999 841	3.13	0.999 990 42
1.77	0.987 691	2.23	0.998 388	2.68	0.999 849	3.14	0.999 991 03
1.79	0.988 641	2.24	0.998 464	2.69	0.999 858	3.15	0.999 991 60
1.80	0.989 091	2.25	0.998 537	2.70	0.999 866	3.16	0.999 992 14
1.81	0.989 525	2.26	0.998 607	2.71	0.999 873	3.17	0.999 992 64
1.82	0.989 943	2.27	0.998 674	2.72	0.999 880	3.18	0.999 993 11
1.83	0.990 347	2.28	0.998 738	2.73	0.999 887	3.19	0.999 993 56
1.84	0.990 736	2.29	0.998 799	2.74	0.999 893	3.20	0.999 993 97
1.85	0.991 111	2.30	0.998 857	2.75	0.999 899	3.21	0.999 994 36
1.86	0.991 472	2.31	0.998 912	2.76	0.999 905	3.22	0.999 994 73
1.87	0.991 821	2.32	0.998 966	2.77	0.999 910	3.23	0.999 995 07
1.88	0.992 156	2.33	0.999 016	2.78	0.999 916	3.24	0.999 995 40
1.89	0.992 479	2.34	0.999 065	2.79	0.999 920	3.25	0.999 995 70
1.90	0.992 790	2.35	0.999 111	2.80	0.999 925	3.26	0.999 995 98
1.91	0.993 090	2.36	0.999 155	2.81	0.999 929	3.27	0.999 996 24
1.92	0.993 378	2.37	0.999 197	2.82	0.999 933	3.28	0.999 996 49
1.93	0.993 656	2.38	0.999 237	2.83	0.999 937	3.29	0.999 996 72
1.94	0.993 923	2.39	0.999 275	2.85	0.999 944	3.30	0.999 996 94
1.95	0.994 179	2.40	0.999 311	2.86	0.999 948	3.31	0.999 997 15
1.96	0.994 426	2.41	0.999 346	2.87	0.999 951	3.32	0.999 997 34
1.97	0.994 664	2.42	0.999 379	2.88	0.999 954	3.33	0.999 997 51
1.98	0.994 892	2.43	0.999 411	2.89	0.999 956	3.34	0.999 997 68
1.99	0.995 111	2.44	0.999 441	2.90	0.999 959	3.35	0.999 997 838
2.00	0.995 322	2.45	0.999 469	2.91	0.999 961	3.36	0.999 997 983
2.01	0.995 525	2.46	0.999 497	2.92	0.999 964	3.37	0.999 998 120
2.02	0.995 719	2.47	0.999 523	2.93	0.999 966	3.38	0.999 998 247
2.03	0.995 906	2.48	0.999 547	2.94	0.999 968	3.39	0.999 998 367
2.04	0.996 086	2.49	0.999 571	2.95	0.999 970	3.40	0.999 998 478
2.05	0.996 258	2.50	0.999 593	2.96	0.999 972	3.41	0.999 998 582
2.06	0.996 423	2.51	0.999 614	2.97	0.999 973	3.42	0.999 998 679
2.07	0.996 582	2.52	0.999 634	2.98	0.999 975	3.43	0.999 998 770
2.08	0.996 734	2.53	0.999 654	2.99	0.999 976	3.44	0.999 998 855
2.09	0.996 880	2.54	0.999 672	3.00	0.999 977 91	3.45	0.999 998 934
2.10	0.997 021	2.55	0.999 689	3.01	0.999 979 26	3.46	0.999 999 008
2.11	0.997 155	2.56	0.999 706	3.02	0.999 980 53	3.47	0.999 999 077
2.12	0.997 284	2.57	0.999 722	3.03	0.999 981 73	3.48	0.999 999 141
2.13	0.997 407	2.58	0.999 736	3.04	0.999 982 86	3.49	0.999 999 201
2.14	0.997 525	2.59	0.999 751	3.05	0.999 983 92	3.50	0.999 999 257
2.15	0.997 639	2.60	0.999 764	3.06	0.999 984 92	3.51	0.999 999 309
2.16	0.997 747	2.61	0.999 777	3.07	0.999 985 86	3.52	0.999 999 358
2.17	0.997 851	2.62	0.999 789	3.08	0.999 986 74	3.53	0.999 999 403
2.18	0.997 951	2.63	0.999 800	3.09	0.999 987 57	3.54	0.999 999 445
2.19	0.998 046	2.64	0.999 811	3.10	0.999 988 35	3.55	0.999 999 485
2.20	0.998 137	2.65	0.999 822	3.11	0.999 989 08	3.56	0.999 999 521
2.21	0.998 224	2.66	0.999 831	3.12	0.999 989 77	3.57	0.999 999 555

w	erf(w)	w	erf(w)	w	erf(w)	w	erf(w)
3.58	0.999 999 587	3.69	0.999 999 820	3.80	0.999 999 923	3.91	0.999 999 968
3.59	0.999 999 617	3.70	0.999 999 833	3.81	0.999 999 929	3.92	0.999 999 970
3.60	0.999 999 644	3.71	0.999 999 845	3.82	0.999 999 934	3.93	0.999 999 973
3.61	0.999 999 670	3.72	0.999 999 857	3.83	0.999 999 939	3.94	0.999 999 975
3.62	0.999 999 694	3.73	0.999 999 867	3.84	0.999 999 944	3.95	0.999 999 977
3.63	0.999 999 716	3.74	0.999 999 877	3.85	0.999 999 948	3.96	0.999 999 979
3.64	0.999 999 736	3.75	0.999 999 886	3.86	0.999 999 952	3.97	0.999 999 980
3.65	0.999 999 756	3.76	0.999 999 895	3.87	0.999 999 956	3.98	0.999 999 982
3.66	0.999 999 773	3.77	0.999 999 903	3.88	0.999 999 959	3.99	0.999 999 983
3.67	0.999 999 790	3.78	0.999 999 910	3.89	0.999 999 962		
3.68	0.999 999 805	3.79	0.999 999 917	3.90	0.999 999 965		

Index

A

Aluminium Metallization 262
Aluminum Etching 146
Ambient control 23
Anisotropy 139
Annealing 223
As Diffusion 193
Atomic mechanisms of Diffusion 168
Au Diffusion 193
Autodoping 58

B

B Diffusion 190
Ball bonding 290
Ball Grid Arrays Package 287
Beam Line System 211
Bi-CMOS Fabrication 310
Bi-CMOS Technology 309
Bipolar IC Technology 307
Building Individual Layers 296
Bulk Defect 17
Buried Insulator 232
Buried Layer Pattern Transfer 53

C

Capacitance-Voltage Plotting (C-V) 199
Chemical Cleaning 32
Chemical Vapor Deposition 42
Clean Room 5
CMOS IC Technology 301
Color Chart 93
Contact Optical Lithography 106
Control System 211
Copper Metallization 261
Crystal Defect 13
Crystal Growth 23
Crystal Manufacturing 20
Crystal Pulling 22
Crystal Structure 10
Cubic Structure 11
CVD 245
Czochralski Technique 20

D

Defectivity 292
Defects 50
Deposition Apparatus 266
Deposition Methods 265
Die Bonding 289
Die Interconnection 288
Diffusion 167
Diffusion in Polysilicon 197
Diffusion Process Properties 178
Diffusion Profiles 172
Diffusion Systems 189
Diffusion Temperature 178
Diffusion Time 178
Diffusivity of Antimony 181
Diffusivity of Arsenic 182
Diffusivity of Boron 183
Diffusivity of Phosphorus 185
DIP 179
Doping Profile of Ion Implant 222
Doping 48
Dry Etching Process 147
Dry Oxidation 79
Dual Diffusion Process 177

E

Electron Beam Lithography 122
Electron Optics 126
Electron Projection Printing 127

Electron Proximity Printing 127
Electron-Matter Interaction 124
Emitter Push Effect 186
Error Function properties 173
Etch Parameters 139
Etch Profile 140
Etch Rate 139
Etching 139
Etching Reactions 159
Etching 31
Evaporation 243
Extrinsic Diffusion 179

F

Fabrication Facilities 317
Fick's Laws of Diffusion 171
Field-Aided Diffusion 188
Fin Etch 314
Finfet 312
Fixed Oxide Charge 93
Float Zone (FZ) Technique 26
Flux 83
Four Point Probe 200
Furnace Annealing 224
Furnace 20

G

Gas source Diffusion 190
Gas System 210
Gattering 32
Gaussian Diffusion 175

H

Hardbake 121
Hermetic Ceramic Packages 280
High Energy Implantation 231
Hybrid or multichip ICs 315

I

IC Fabrication 316
Implanted Silicides and Polysilicon 230
Impurity Effect on the Oxide Rate 87

Impurity Redistribution 195
Inductive coupled Plasma Etching (ICP) 153
Ingot slicing 28
Ingot trimming 28
Integrated Circuit Package 284
Interface-trapped charges 91
Interstitial Diffusion 169
Intrinsic Diffusion 169
Ion Beam Lithography 127
Ion Implant Stop Mechanism 213
Ion Implantation 208
Ion Implanter 209
Ion Source 211

L

Lateral Diffusion 196
Lateral Diffusion Effects 179
Liftoff 157
Liftoff Process 268
Limited Source Diffusion 175
Line Defect 15
Liquid Phase Epitaxy 39
Liquid Source Diffusion 192
Lithography 103
Locos Methods 254
Low Energy Implantation 229
Low temperature epitaxy 60

M

Masks 112
Mass Analyzer 211
Metallization 257
Metallization Patterning 268
Metallization Processes 265
Mettalization Choices 260
Microscopic Growth 52
Miniaturizing VLSI Circuits 298
Mobile Ionic Charge 91
Molecular Beam Epitaxy 61

Index 335

Monolithic ICs 315
Moore's Law 3
Multilevel Metallization 269

N

NMOS IC Technology 298
N-Well 302

O

Ohmic Contacts 259
Optical Lithography 105
Oxidation Growth and Kinetics 78
Oxidation Techniques 92
Oxidation 75
Oxide Charges 90
Oxide Furnaces 95
Oxide Masking 193
Oxide Property 89
Oxide Thickness Measurement 92
Oxide Thickness 87
Oxide Trapped Charge 92

P

Package Types 280
Packaging 279
Packaging Design Considerations 283
Particle-Based Lithography 122
Pattern inspection 121
Pattern Transfer 119
Phase Shifting Mask 115
Photomask Fabrication 114
Photoresist 116
Photoresist Stripping 121
Physical Vapor Deposition 242
Physical Vapor Deposition 265
Physical vapor deposition 61
Planarized Metallization 271
Plasma Chemical Etching Process 150
Plasma Etching Process 147
Plastic Packages 280
Point Defect 13

Polishing 31
Polysilicon 256
Post Acceleration 213
PQFP 279
Process Flow 321
Projection Optical Lithography 107
Proximity Optical Lithography 106
Proximity Printing 131
P-Well 305

Q

QFN 286
Quad Flat Packages 217

R

Range and Straggle of Ion Implant 217
Rapid Thermal Annealing 226
Reactive Ion Etching (RIE) Process 152
Reactors 49
Recess Etch 314
Reflectivity of Metal Film 274
Resists 126

S

Sb Diffusion 193
Secondary Ion Mass Spectroscopy 201
Selective Epitaxy 59
Selectivity 140
Semiconductor Material 8
Shallow Junction Formation 228
Silicon Dioxide 249
Silicon Dioxide Etching 145
Silicon Etching 143
Silicon Nitride 146
Silicon Nitride 253
Silicon on Insulator 67
Silicon on Sapphire 67
Silicon on SiO_2 68
Silicon Properties 18
Silicon Purification 18
Silicon shaping 27

Small-Outline Package 279
Softbake 120
Solid Solubility 178
Solid Source Diffusion 191
Spreading Resistance Probe 201
Sputter Etching Process 151
Sputtering 244
Sputtering 61
Staining 198
Step Coverage and Reflow 250
Stress of Metal Film 273
Substitutional Diffusion 168
Surface Defect 16
Surface-Mount Package 284

T

Theortical Treatment 44
Thickness of Masking 220
Thickness of Metal Film 271
Through-hole package 284
Tilted Ion Beam 229
Twin Tub Process 306

U

Ultra Violet Lithography 129
Uniform Defect Densities 292
Uniformity of Metal Film 273
Uniformity 52

V

Vacuum System 211
Vapour phase epitaxy 39
VLSI assembly Technologies 287
VLSI Generation 2
Volume Defect 17

W

Wafer Processing 32
Wedge Bonding 294
Wet Etching Process 141
Wet Oxidation 80
Wire Bonding 289

X

XRay Lithography 130
X-Ray Masks 132
X-Ray Resist 131
X-Ray Sources 132

Y

Yield 291